An Economic Analysis of Conflicts

Nicos Christodoulakis

An Economic Analysis of Conflicts

With an Application to the Greek Civil War 1946-1949

Springer

Nicos Christodoulakis
Department of International & European Economics Studies
Athens University of Economics & Business (AUEB)
Athens, Greece

ISBN 978-3-319-32260-5 ISBN 978-3-319-32261-2 (eBook)
DOI 10.1007/978-3-319-32261-2

Library of Congress Control Number: 2016946028

Printed on acid-free paper

This Springer imprint is published by Springer Nature
The registered company is Springer International Publishing AG Switzerland

Preface

This book has been a product of serendipity. In 2013, I was doing research on the economic and currency crisis that hit Greece in the interwar period and then started to look at economic developments in the 1940s. The lack of statistical data covering the period of Axis Occupation (1941–1945) and the subsequent Civil War (1946–1949) led me to search for studies and reports prepared by the British and US missions in Greece at that time.

Eventually, I visited the US National Archives in Washington DC and asked for some declassified reports that were prepared in the late 1940s by JUSMAG, the Joint US Military Advisory Group to Greece. The material consisted mostly of trivial reports on the economic situation, except that some documents included a systematic recording of all the battles that took place during the Civil War. There were monthly data covering battle casualties and army formations of the guerrillas and the Greek National Army, as well as their geographical dispersion for the period stretching from January 1948 to December 1949.

I couldn't believe my luck. As far as I know, this is one of the very few systematic recordings of a civil war that is available to date, even if recent intrastate conflicts worldwide are taken into account.

Upon returning to Greece, I set about to complete the data series backwards and thus obtain a full record since the start of the Civil War in 1946. To that effect, all available military archives of the Greek Army and a number of reports released by the Communist Party of Greece were classified in a pattern similar to that of the US series. In this way, a complete monthly battle database was obtained for 1946–1947. Additional data covering persecutions of political opponents and the destruction of productive activities during the conflict were also compiled from relevant sources.

Reading the prolific academic literature on civil wars, either contemporary or past ones, I was impressed by the scarcity of quantitative and systematic evidence available on the evolution of specific civil wars over time. The vast majority of empirical work is based on cross-country data. Usually, these series cover total battle casualties and aggregate socio-economic conditions and are subsequently used in cross-estimates to test alternative hypotheses on intrastate conflicts. Apart from implicitly making the assumption that countries respond to such conflicts in a similar manner, cross-section analysis is not capable of capturing the dynamics of a civil war. The particular patterns each conflict has followed, or the specific conditions under which it broke out or was terminated, are not possible to analyse.

For that reason, I thought that it would perhaps be useful to concentrate on how a civil war can be modelled in a single country. The task involved the derivation of appropriate combat models or a substantial modification of existing ones, as well as setting up a number of econometric procedures so as to properly estimate battle interactions. Then the battle time series available on the Greek conflict were used to test the properties of the dynamic process. Occasionally, some of the model predictions are compared by using the Correlates of Civil War (COCW) database. I hope that the outcome will contribute both to theoretical aspects of modelling conflict situations and to a detailed description and understanding of the dynamic patterns observed in the Greek Civil War. Although motivated by the examination of a civil war in a particular country, this book hopes to offer some analytical tools that are useful in improving the analysis on—and the understanding of—many others.

Athens, Greece
April 2016

Nicos Christodoulakis

Acknowledgements

I am deeply obliged to Demetrios Varakis and Christos Triantopoulos for the meticulous compilation of data from Greek military archives. This book would not have been possible without the help, advice and comments received in several seminars in which earlier parts of this research were presented. I thank the participants in seminars held in the Bank of Greece, the London School of Economics, the Athens University of Economics, the Yale University and the Independent Archives of Social History in Athens (ASKI) for their useful remarks and constructive criticism. I also wish to thank the Journal of Defence and Peace Economics for publishing a substantial part of my research on the Greek Civil War, and especially the Editor, Professor Christos Kollias, for his encouragement and suggestions throughout the early research project.

Chapter 2, Sects. 3.7, 4.7, 6.2, 6.3, 7.6, 8.5, 8.7, 9.2, 10.1, 10.2, and Data Appendix are derived, in part, from two articles published in the Journal of Defence and Peace Economics on 28 Jan 2015 and 02 Mar 2015, respectively, available online: http://dx.doi.org/10.1080/10242694.2014.1000010, http://dx.doi.org/10.1080/10242694.2015.1016297

Abbreviations

COCW	Correlates of Civil War
DAG	Democratic Army of Greece
GCW	Greek Civil War
GDP	Gross Domestic Product
GNA	Greek National Army
HQ	Headquarters
JUSMAG	Joint US Military Advisory Group to Greece
KKE	The Communist Party of Greece
SIPRI	Stockholm International Peace Research Institute
UK	United Kingdom
US	United States of America
USSR	The Soviet Union
WW1	First World War
WW2	Second World War

Mathematical

i.i.d.	independent and identically distributed
LHS	left-hand side
OLS	ordinary least squares
RHS	right-hand side
SUR	seemingly unrelated regression
w.r.t.	with respect to

Contents

List of Figures

List of Tables

Chapter 1
Introduction: Economic Analysis and Civil Wars

The Introduction examines the main channels that interconnect economics and civil wars, both in influencing the causes of conflict and affecting its outcome. It also examines how economic theory and econometric analysis can help model the dynamics of conflict and thus estimate its patterns and consequences.

1.1 The Economic Effects of Conflict

The two main channels interconnecting economics and civil wars are concerned with the causes of conflict and its effects on the economy. The former is a matter of open and fierce debate as to whether economic considerations are the dominant motivation for a group to enter a conflict, or there are such widespread grievances that eventually bring the strife to a costly bloodshed. The key rival theories are examined in the next section and are later tested in various applications throughout the book.

The effects of a civil war on the national economy are less controversial, and are discussed below. Regarding the economic legacy of a domestic conflict, Collier and Hoeffler (2007) estimate that a civil war results in a loss totalling between 90 % and 105 % of a year's GDP, which is materialized either directly or indirectly. The toll may further rise by a series of growth-inflicting externalities, ranging from the disruption of markets to the collapse of international trade and the deterrence of foreign investment. A non-exhaustive list of economic detriments induced by a civil war includes the following:

(a) The direct costs of conflict come from the destruction of the factors of production. People are killed or seriously injured, thus they are incapacitated from further participating in the country's labor force and human capital formation. The capital stock used in, or supporting, production is extensively destroyed if hostilities escalate and last for a long period of time. Factory equipment,

N. Christodoulakis, *An Economic Analysis of Conflicts*,
DOI 10.1007/978-3-319-32261-2_1

infrastructures, as well as the livestock in rural areas, are destroyed by fighting or looted by rival factions. In many cases, the loss of human and physical capital is accompanied by a loss in technology that has a further adverse effect on the level and quality of production.

(b) The lack of access to foreign markets because of transportation problems in the areas of military operations or the imposition of international sanctions and embargos against the country. In some cases, the process of a country being assimilated or integrated in a regional or international organization may be seriously delayed or even altogether abandoned.

(c) The deterrence of foreign investment since a protracted conflict causes political instability and undermines institutions and credibility. In several cases, the investment on human capital formation may also cancelled if people migrate to other countries either in order to try to avoid persecution at home or seek better opportunities elsewhere.

(d) The drop in production owing to threats against property rights by actions of output expropriation and/or sabotage. The role of sabotage against infrastructure and major production units has long been recognized as being a critical factor to the outcome of a civil war.[1]

(e) The misallocation of resources, as the government shifts budget priorities to military and security spending in order to support troops fighting against the insurgents or alleviate the recurrence of conflict. As a result, critical areas of infrastructure building, social policy and technology dissemination remain underfinanced and growth is stunted for many years.

(f) The proliferation of rent-seeking activities against productive ones. A protracted conflict leads to the formation of vested interests in the sphere of politics and the economy, which try to perpetuate their privileges long after the conflict is ended. For example, a shift to armaments spending is likely to continue afterwards, as the role of the military in winning the civil war establishes a permanent influence on public decision-making. A host of distortions associated with increased military spending, from over-militarization of society to outright corruption, are likely to follow and undercut the prospects of economic development.

(g) Fueling further discontent: A reinforcing mechanism between low growth and higher propensity to conflict is frequently put in motion. Low growth precipitates income inequality, poverty and exclusion, thus feeding grievances and also enabling the motivation to participate by lowering income foregone.

[1]Instructing modern revolutionaries, Che Guevara (1961, Ch. 3) advocated that "it is absolutely correct and advisable to carry out sabotage against a power plant. . . . [T]his is entirely justified by the paralysis of the life of the region."

1.2 The Conflict of Economics: Rival Explanations of Causes

In attempting to explain why a civil war erupts in the first place, academic literature focuses on the root-causes that accumulate over time and either provide incentives for, or amount to a justification of, participating in a conflict. The framework of analysis assumes that individuals or organized groups consider alternative options for improving the prospects of their own utility. Evaluating the options, they decide whether the use of force against the *status quo* is likely to confer benefits or is doomed to fail and the outcome will be inferior to the present state of things. A number of sociopolitical, economic and natural constraints determine the viability and payoffs of such an endeavour, thus influencing the decision on whether to use force, the duration of conflict, and the odds of emerging on the winning or the losing side.

The root-causes may be such that affect each individual, or organized group, in pursuing their own interests, and in this case a conflict is seen as a calculated outcome of rational choice. According to an influential paper by Collier and Hoeffler (2001), an individual participates in an insurgence if expected gains outweigh the costs of engagement, including the benefits foregone by abandoning current activities. Theories of rational choice are further enriched by assuming that certain characteristics of expected gains and costs are influenced by social, political or geography factors.

In an authoritative review on civil wars, Sambanis (2002) criticized the opportunity-cost model and suggested that the accumulation of grievances in a society is more likely to exasperate certain groups that subsequently choose to resort to rebellion. Collective or group grievances may have reached such levels that the overall socioeconomic structure is intolerable or unsustainable. This causes such widespread exasperation that some groups find the opportunity to organize an agenda of confrontation and people opt for the conflict as the only available means to improve the overall, as well as their own, situation.

As noted by Lu and Thies (2011), grievance precedes opportunity and denotes just a motivation to rebel, which may be activated only if the opportunity cost of fighting becomes very low whenever economic conditions deteriorate or exclusion politics dominate. The mechanism through which grievances feed instability and political animosity is explained by arguing that:

> ...people become aggrieved when they feel that they are disadvantaged relative to others within a polity or relative to what they perceive as justifiable. [These fuel] the hatred that rebels feel toward the government and the ensuing violence that comes when they fight against it.

Grievance theory is not new, and in fact has been adopted by several revolutionary leaders in their advocacy of conflict when all other means have. One of the most outspoken on using grievances to justify a rebellion was Lenin (1929, p. 289), who openly asserted that

> ...the oppressed masses never won in a struggle involving losses of life, unless they were previously put in a state of despair due to long lasting hardships and acute crises.

Such theories are echoing the 'calculus of advantage' proposed by Benthamite utilitarianism as the mechanism that makes collective action worth pursuing under specific conditions. In practice, however, grievances thrive with poverty and political exclusion, but still may fall short of spurring a civil war. Greed theory may also be at odds with explaining costly conflicts that seriously damage both adversaries, irrespective of the outcome. The eruption of a civil war is more likely when the opportunity cost of rebellion is reduced or the striking capability of the interested group is enhanced due—among other reasons—to weak enforcement institutions, ineffective repression or some historical idiosyncrasy, such as a previous war or a national tragedy.

Fearon and Laitin (2003) argue that violence is escalated only if repression is poorly enforced, thus effectively reducing engagement costs and increasing expected payoffs for insurgents. Frequently the escalation of repression leads to larger-scale hostilities instead of suppressing them. In contrast, democratic institutions may prove efficient in alleviating grievances and thus prevent bloodletting.

The gap between greed and grievance theories was finally closed when the main authors of the seemingly competing approaches applied them jointly in order to study the occurrence and severity of modern civil wars; see Collier and Sambanis (2005). In his foreword to the two-volume publication by the World Bank, Ian Bannon—Manager of the Bank's Conflict Prevention and Reconstruction Unit—did not miss the opportunity to note that

> [a]n important additional benefit of this research is that it may lay to rest the *"greed versus grievance"* caricature. The research presented here makes it clear that greed and grievances should not be seen as competing explanations of conflict—they are often shades of the same problem.

1.3 Empirical Relevance

To establish empirical relevance, the above theories are tested by using cross-country or regional panel data to examine correlations between severity, duration and spread of civil wars with various explanatory variables, ranging from social indices and economic conditions to national idiosyncracies and geography. An indicative selection of findings is described.

According to Brückner and Ciccone (2007), low growth of per capita income makes the eruption of civil war more likely, albeit only in non-democracies. Pooling data from 39 Sub-Saharan African countries for the period 1980–2003, it is found that a 5 % fall in economic growth increases the likelihood of civil war by around 8 % in non-democratic countries. This effect is significantly weaker in countries with democratic institutions, where no significant impact of income growth on civil war is established. The apparent explanation is that in democratic countries it is more likely to see the implementation of policies that improve citizen participation in the distribution of income, thus increasing the opportunity cost of taking part in a conflict.

Blattman and Miguel (2009) explore various root-causes theories and conclude that slow economic growth and favouring geographic conditions are found to be robustly linked to civil wars. Moreover, as pointed by Regan and Norton (2005), grievance is not automatically transformed into rebellion, unless party elites are capable of collective mobilization and of ensuring protection for those participating. Besley and Persson (2008) have shown that more consensual political institutions reduce the incidence of conflict, as the distribution of more public goods diminishes the value of unilateral appropriation by the party in power.

There are two potential problems with the above approach. First is that by conducting cross-country estimates, one makes the implicit assumption that different countries observed in different epochs respond in a similar way regarding the possibility of civil war and its ensuing dynamics. Although country-specific or time-specific dummies are used in the estimation to factor in local idiosyncrasies, this only affects the degree to which each country responds to the correlates of war, without challenging whether the mechanism may apply in an entirely different manner across states.

The second—and more crucial- issue is that, among the multitude of conflict—related factors, a distinction should be made with regards to the causation effects and the facilitation mechanism. Causation factors include those which affect the motivation to enter a conflict, and may be derived either from material considerations of costs and opportunity or from other value systems, such as ideology, ethnicity or religion. These may or may not be sufficient to lead to a definite decision to enter a conflict, and in fact they will continue to apply no matter if the conflict actually ensues.

The other set includes the factors that facilitate a conflict *after* it has been started, and are otherwise irrelevant. For example, certain patterns of terrain may enable or disable combatant forces, but is ridiculous to assume that they *motivate* the relevant decision. Newman (2014, Ch. 2) pointedly warns that

> ...broad approaches to explaining civil wars often neglect the fact that armed conflicts evolve after they begin; the dynamics that explain their onset may not necessarily be the same as those that explain their ongoing nature. Indeed very often, new dynamics emerge.

1.4 Conflict as an Institutional Failure

One reason for a conflict to ensue may be due to the failure of adversaries to negotiate effectively on the presumption that termination is preferable over bloodletting. This may occur, for example, because potential agreements lack an enforcement technology and the other side believes that they will be overturned as soon as a ceasefire is declared.

This happens whenever there is uncertainty about the true motives of the other side because of incomplete and asymmetric information, and there are several reasons leading to such outcomes: Skaperdas (2008) argues that the ability to enforce negotiated contracts between competing groups is weakened by various factors ranging from geography and ethnicity to external intervention. This makes the option of war as a means of appropriating power look more appealing, even if the cost of engagement is multiplied.

Another strand of literature attributes the occurrence and longevity of civil wars to incomplete and asymmetric information that distorts the payoffs anticipated by the adversaries. In a notable contribution, Acemoglu and Wolitzky (2014) argue that incomplete information about a rival's intentions may lead one side to interpret noisy signals as pretentious tactics by the other that are only designed to secure an advantage, and thus opt to respond in a similar manner.[2] The outcome is that each side maximizes its own aggression, leading to "conflict spirals."

The failure to negotiate effectively, is owed either to uncertainty about ulterior motives or simply because agreements lack an enforcement technology. Cunningham (2013) argues that negotiations by proxy may fail because 'in many cases mediators and others may have interests beyond just ending the fighting' and it may be the case that 'international actors are unwilling to intervene to the degree necessary to forcibly disarm veto players'. Uncertainties multiply in the absence of an incontestable leader, and it is well-known that the closer the electoral scores of the partners of a coalition, the more competitive they are with each other; for an analysis see Tsebelis (1988). But even when leaderships truly seek a compromise, their efforts may be impeded by the accumulation of aggression exercised at a decentralized level. As analyzed by Kalyvas (2006, Ch. 9), the spreading of violence at a micro-level proved to be such a strong fertilizer of the conflict, that it gave the hardliners of both sides all the excuses they needed to interpret the actions as signals of mounting aggression by their opponent.

1.5 Conflict Modelling

A variety of approaches is adopted with regard to the modelling of conflict situations. Perhaps the most comprehensive review of conflict modelling is given by Lichbach (1992), who groups 200 scholarly contributions into two broad categories: those falling into the rational-choice optimizing framework and the stochastic models without the presumption of rationality. The former set is employed by the root-causes literature by testing—usually by cross country or regional panel data—the intensity of conflict against various explanatory variables.

The second set studies the dynamics of either single-equation or interaction models, usually variants of Lotka-Volterra systems that were originally devised to

[2]This kind of risk applies in conflicts where the opponents truly seek for a face-saving compromise. In cases where an aggressive ideology is dominant, compromising signals are likely to be ignored deliberately.

study species interactions. Later such models were employed for studying the interactions generated by the class-struggle (e.g. Goodwin 1967), the arms race (e.g. Richardson 1960), political rivalry (e.g. Francisco 1996), riots (e.g. Burbeck et al. 1978) or outright revolution (e.g. Tsebelis and Sprague 1989).

1.5.1 The Two-Numbers Problem

In a NATO Conference on combat modeling that took place in 1967, a senior Defense analyst complained that in most cases only two numbers (i.e. on the terminal casualties of the two adversaries) are available to describe the outcome of a war, thus rendering model estimation impossible.[3] As a matter of fact, the majority of contemporary studies on civil wars continue to use just two numbers related to battle losses per case. Then by pooling many wars together, they try to obtain correlates with socio-economic and political variables. In this way, each civil war is treated as a "black-box", without any particular analysis regarding the idiosyncrasies and dynamics of the specific process.

1.5.2 Enter Dynamics

In contrast, studying a nation-case in detail helps to study the phases and the extent of the conflict in a deeper sense. Furthermore, it will also help to improve the understanding of the way the correlates of war interplay with other conflict characteristics, without making the slippery assumption that the country is similar to others. In particular, a dynamic analysis allows the examination of the following:

1. *Conflict traps:* By applying dynamic analysis on a specific time-series one can calculate any existing conflict equilibrium between the two adversaries. Then, the stability properties are determined and the reaction to various shocks can be described.
2. *Losses:* Human losses are recorded in each period of time. This allows for a more precise assessment of the destruction of the factors of production and the loss of output due to the civil war.
3. *Counterfactual analysis:* Counterfactual scenarios can be studied, by assuming that hostilities terminate at an earlier period. The output loss that occurs afterwards and up to the point of actual termination may be considered as the price of intransigence and is taken as a measure of the negotiation failure.
4. *Phases of conflict:* Employing a long series of data from a specific war, one can better examine how critical some factors are in shaping the outcome of the

[3]The remark was made by R. L. Helmbold in Munich and quoted by Wallis (1968).

conflict. For example, terrain, funding from ethnic Diasporas, and the provision of military equipment by foreign powers may affect the course of the war at a specific phase and this is properly detected only by analyzing a time series of the conflict in question.

1.6 Conflict Definitions

A conflict is considered to be a civil war when certain characteristics are in place, such as the following:

First, the extent and severity of the conflict is considered. To distinguish small-scale violence from large-scale and organized hostilities, it has been established that a conflict is recorded as a civil war only if conflict-related killings exceed 1000 per year.

Second, a civil war differs from a situation of widespread lawlessness in the sense that it is the organized and centrally coordinated action of a certain group serving specific interests against another group or set of interests. Situations characterized by atomized violence against property or people do not necessarily fall in the category of a civil war. In fact, most of the organized conflicts in modern times rush to denounce sporadic acts of violence and try to attract maximum force in their ranks. For communist-led rebellions, such a prerogative was critical for the chances of victory to be enhanced. According to Lenin (1965),

> . . . an uprising cannot assume *the old form of individual acts* restricted to a very short time and to a very small area. It is absolutely natural and inevitable that the uprising should assume the higher and more complex form of a prolonged civil war embracing the whole country, i.e., *an armed struggle between two sections of the people;* (emphasis added).

Third, a central agenda is usually adopted by the insurgents, both in order to inspire prospective participants and to provide a political, ideological or religious framework to discipline the fighting force. Mao (1937), perhaps the most successful guerrilla leader of the twentieth century, insisted that

> Without a political goal, guerrilla warfare must fail, as it must, if its political objectives do not coincide with the aspirations of the people and their sympathy, co-operation, and assistance cannot be gained.

Another constant inspiration of several organized conflicts in recent decades has been Che Guevara, who, on the same subject, left no doubt about the need of a socially appealing agenda, advising that:

> . . . the guerrilla fighter is a social reformer, that he takes up arms responding to the angry protest of the people against their oppressors, and that he fights in order to change the social system. . .

But even less romanticised figures were equally adamant in claiming an agenda far beyond their narrow interests, in order to appeal to a wider audience for purposes of recruitment and support in general. For example, Foday Sankoh, the notorious warlord of the civil war in Sierra Leone in the 1990s, defended his actions in the name of the people by saying that[4]:

> The agents of colonial masters called politicians made our people poor. This is why we have been fighting. We are not against our people. We are against a rotten system. [. . .] We have been misunderstood.

Of course, in real-life conflicts ideology may be just a convenient veil for the otherwise clear objective of direct seizure of power and the resulting privileges. Newman (2014, p. 64) aptly notes that

> [s]ome such conflicts do not. . . reflect a struggle between competing ideologies, but rather a struggle over control of the state and the material benefits it bestows to political elites and their supporters.

According to SIPRI, civil war is a major armed conflict, defined "as a prolonged combat between the military forces of two or more governments, or of one government and at least one organized armed group, and incurring the battle-related deaths of at least 1000 people during the entire conflict. A conflict location is the territory of a state."

Using the above definition, Newman (2014, p. 62) considers that the following processes fall within the definition of a civil war:

1. An armed conflict within recognized national borders and with the key protagonists being local.
2. The exercise of systematic and organized violence between organized groups, whether supporting or opposing the government.
3. Significant human toll on more than one sides that take part in the conflict.
4. The main parties involved in the conflict coalesce around political, social, or ideological agendas and processes.

1.7 Book Structure

The rest of the book deals with the problem of modelling such situations and the description of their dynamics.

Chapter 2 provides a brief on the origins and main episodes of the Greek Civil War between the Communist Party of Greece and the centre-right government during 1946–1949. It also describes the data series that are subsequently used to obtain estimates of battle dynamics and determine the separate Phases of conflict.

Chapter 3 reviews the original Collier-Hoeffler model that tries to explain the onset of civil wars by employing an opportunity-cost framework. Using regional data for Greece, the hypothesis that political discontent thrives on the multiplication/growth of contemporaneous grievances or the class structure is tested and found to be irrelevant for explaining social rivalries. In contrast, some

[4] Recorded in http://news.bbc.co.uk/2/hi/africa/629753.stm, Feb, 4, 2000.

old-entrenched grievances stemming from pre-war economic and social problems are found to be more likely to explain opposition to the government.

Chapter 4 sets up a modified version of the Collier-Hoeffler model and examines two different stages in a civil war; one before and one after engagement has been decided. In the period preceding the conflict, a number of socio-political factors are found to be critical on whether an insurgency is initiated in the first place. Societies that use to be more inclusive and less repressive are found to be less prone to engaging in a civil war. In contrast, in exclusive social structures only a high degree of repression may thwart a conflict and only temporarily, and the empirical analysis in Greece shows that repression has been counterproductive.

Chapter 5 introduces the main two families of dynamic conflict models: The combat models of mutual attrition derived by Lanchester, and the Lotka-Volterra models of species interactions. The existence and stability of conflict equilibria are examined and a number of shortcomings are discussed. A modified version is set up so as to capture certain important conflict characteristics and points to the existence of conflict traps that are so common in civil wars.

Chapter 6 describes an econometric methodology on how dynamic models of conflict can be estimated without incurring the problem of co-determination between variables. The method is then applied to the theoretical models derived in Chap. 5, by employing the time-series described in Chap. 2.

Chapter 7 extends the models used in Chaps. 4 and 5 towards a unified framework of conflict, in which an optimal battle plan is devised in each period under the constraint of the army accumulation process. Each side chooses the battle effort to maximize a discounted sum of intertemporal battle objectives. The system is found to have unique saddle-path equilibrium, and its properties are shown to be in line with some stylized facts observed in civil wars. Applications are displayed by using data from conflicts worldwide, as well as from the Greek civil war.

Chapter 8 examines the effects of a civil war on the country's output. First, an endogenous growth model is set up to show how a civil war leads to a higher allocation of public revenues towards the protection of property rights and also to higher taxation, thus lowering economic prospects. An exogenous growth model is also described and used to determine how output is affected by the loss or destruction of production factors. The model is then used to obtain an estimate of GDP losses in Greece for the period 1946–1960. The same model can be used for a counterfactual analysis to assess how much of this cost could have been avoided, had the conflict terminated at an earlier date.

Chapter 9 introduces an alternative framework for estimating GDP losses by comparing the war-afflicted country with other conflict-free economies. In this way, a 'synthetic country' is obtained and the hypothetical GDP represents potential output in the absence of conflict. First five Nazi-occupied European countries are compared with neutral ones to determine the war effect. Then a synthetic Greece is obtained by assimilating its economy with the post-war boom in Western Europe and comparing it to actual GDP. The gap is a measure of GDP forgone, and is found to be somewhat higher than the estimate obtained through the production function approach.

Chapter 10 gives a critical appraisal of how the economic models can be used to broaden our understanding of conflicts and then considers civil wars as major country failures. Finally definitions and sources of time-series are described in the Data Appendix.

References

Acemoglu D, Wolitzky A (2014) Cycles of conflict: an economic model. Am Econ Rev 104 (4):1350–1367

Besley T, Persson T (2008) The incidence of Civil War: theory and evidence. NBER Working Paper 14585. http://www.nber.org/papers/w14585

Blattman C, Miguel E (2009) Civil War. NBER Working Paper 14801. http://www.nber.org/papers/w14801

Brückner M, Ciccone A (2007) Growth, democracy, and Civil War. Federal Reserve Bank of St. Louis, St. Louis, http://ideas.repec.org/p/cpr/ceprdp/6568.html

Burbeck S, Raine W, Stark A (1978) The dynamics of riot growth: an epidemiological approach. J Math Sociology 6(1):1. doi:10.1080/0022250X.1978.9989878

Collier P, Hoeffler A (2001) Greed and grievance in civil war. World Bank, Policy Research Working Paper No. 2355, October

Collier P, Hoeffler A (2007) Civil war. In Sandler T, Hartley K (eds) Handbook of defense economics, vol. 2. Elsevier, pp 712–737. https://openknowledge.worldbank.org/handle/10986/7437 License: CC BY 3.0 IGO

Collier P, Sambanis N (2005) Understanding civil war: evidence and analysis, Volume 1. Africa. World Bank, Washington, DC

Cunningham D (2013) Who should be at the table?: Veto players and peace processes in Civil War. Penn J Law Int Aff 2(1):38–47

Fearon J, Laitin D (2003) Ethnicity, insurgency and civil war. Am Polit Sci Rev 97(1):75–90

Francisco R (1996) Coercion and protest: an empirical test in two democratic states. Am J Polit Sci 40(4):1179–1204

Goodwin R (1967) A growth cycle. In: Feinstein CH (ed) Socialism, capitalism and economic growth. Cambridge University Press, Cambridge

Guevara "Che" E (1961) Guerrilla warfare. https://armexploitednations.net/lib/-/guerrillawarfare.pdf

Kalyvas S (2006) The logic of violence in civil war. Cambridge University Press, New York

Lenin VI (1929) Preparing for revolt. Modern Books, London

Lenin VI (1965) "Guerrilla warfare" collected works, vol 11. Progress Publishers, Moscow, pp 213–223

Lichbach M (1992) Nobody cites nobody else: mathematical models of domestic political conflict. Def Econ 3:341–357

Lu L, Thies CG (2011) Economic grievance and the severity of civil war. Civil Wars 13 (3):215–231. doi:10.1080/13698249.2011.599997

Mao Z (1937) On Guerrilla warfare. Selected works, vol. IX, Ch. 1. http://dx.doi.org/10.1080/13698249.2011.599997

Newman E (2014) Understanding civil wars: continuity and change in intrastate conflicts. Routledge, London

Regan P, Norton D (2005) Greed, grievance, and mobilization in civil wars. J Confl Resolut 49:319–336

Richardson L (1960) Arms and insecurity. Homewood, Pittsburgh (First edition 1919)

Sambanis N (2002) A review of recent advances and future directions in the quantitative literature on civil war. Defence Peace Econ 13(3):215–243

Skaperdas S (2008) An economic approach to analyzing civil wars. Econ Gov 9:25–44. doi:10.1007/s10101-007-0043-2
Tsebelis G (1988) When do allies become rivals? Compar Polit 20(2):233–240
Tsebelis G, Sprague J (1989) Coercion and revolution: variations on a predator-prey model. Math Comput Model 12(4–5):547–559
Wallis PR (1968) Recent developments in Lanchester theory. Oper Res 19(2):191–195, http://www.jstor.org/stable/3008650

Chapter 2
The Greek Civil War 1946–1949: Main Events and Data

The origins and main events of the civil war that took place in Greece during 1946–1949 are briefly described. A new set of battle data compiled from US military archives, Greek Army sources and Communist Party documents is presented and used to provide a quantitative analysis of the armed confrontation. A statistical analysis indicates that there have been two separate phases in the conflict, and the factors that led to the escalation of hostilities are examined.

2.1 The Greek Conflict

2.1.1 Origins

The origins of social discontent and political strife in Greece can be traced back to the interwar period, and the main events are listed in a chronological order at the end of the book (Table 11.5, Appendix). During the interwar period the country moved—like a pendulum—from victory in the First World War (WW1) to defeat in the Asia Minor campaign in 1922, and then again from achieving full membership of the Gold Exchange System into debt defaulting and political instability. In the aftermath of defeat in 1922, Greek society was apoplectic over the national humiliation, while the state was overstretched by the arrival of over a million ethnic refugees. Unemployment ran very high, and the economy was ruined by hyperinflation as a result of a bloated war debt, domestic defaults, and serial currency collapses. The socio-political system became very divisive amid accusations of treason and military coups, while the radicalization of urban and rural populations by the refugees and the empowerment of the Communist Party of Greece (KKE) led to serious clashes with authorities.

Then, a period of political stability settled in, and several steps were taken towards restructuring and modernizing the economy. During 1928–1932, Greece

N. Christodoulakis, *An Economic Analysis of Conflicts*,
DOI 10.1007/978-3-319-32261-2_2

joined the Gold Exchange System and tried to regain part of its role in the European scene. But soon afterwards Greece was hit by the international crisis, lost its foreign exchange reserves, abandoned the Golden peg and defaulted on its external debt.[1] The consequences came fast and were vast. Initially, and as a result of the steep currency devaluation, there was a revival of industrial production, which nonetheless proved to be neither adequate to compensate for the impoverishment of the agricultural sector, nor capable of stopping the destruction of small enterprises and the process of rapid proletarianization in the cities.

The country fell into chaos and political disintegration. Four consecutive elections (held in 1932, 1933, 1935 and 1936), failed to produce a stable government that would manage the economic situation. After each electoral round political tensions were escalated with many atrocities taking place, including an assassination attempt against the crisis Prime Minister, one election boycott, and four military *coups d' etat*. Two of them were staged by the Party of Liberals, and their leaders were summarily executed afterwards. The third coup was royalist and managed to restore the monarchy through an allegedly rigged referendum. As unemployment continued to soar and social clashes intensified, a right-wing dictatorship was finally imposed by yet another coup in 1936. During the regime, republicans, trade unionists, and communist sympathizers were severely persecuted; those in prison were later handed over to the Axis powers that occupied Greece in 1941.[2]

2.1.2 Occupation

Greece was devastated during the triple Occupation by Germany, Italy and Bulgaria. The population was decimated either by mass reprisals or starvation. The Red Cross has estimated that, from 1941 until 1943, food shortages alone caused the Greek population to decrease by at least 300,000.[3] Infrastructure, houses and social services were destroyed at a massive scale, as documented by Doxiadis (1946).

The obvious economic cost was the dramatic fall in output. Greek GDP in 1945 was down by 64 % from its level in 1939, the year before the war started. According to the Bank of Greece (1947), the Drachma was collapsing so fast that on Liberation day in October 1944, the exchange rate to the British sovereign had skyrocketed by more than 1.6 billion times relative to its pre-war level. Hyperinflation had ruined

[1]An analysis of the 1932 crisis in Greece and its dire consequences on the economy is given by Christodoulakis (2013).

[2]For an account of police persecutions of republicans and left-wing sympathizers, see Close (1986) and Pelt (2014).

[3]As compared to the population the country would have had in the absence of war; see Mazower (1994, p. 67).

any perceivable form of market exchange; as a result, barter and black market became normal practices. Moreover, the Axis authorities had forcibly obtained a loan of US$ 228 million in 1944 prices from the Bank of Greece to finance emergency occupation costs.[4]

During Occupation, a broad-based resistance movement was organized by the Communist Party throughout the country, pledging the transformation of post-war Greece into a People's Republic. On the eve of Liberation in October 1945, the Communist Party participated in the National Government together with the conservative and liberal parties, while keeping large parts of the countryside and several cities under its control.

Domestic political rivalries were little subdued and were soon exacerbated again, leading to open conflict. Britain was overseeing the liberation process, and insisted on the disarming of resistance fighters and the reinstatement of the King, a condition fiercely opposed by the Left. In December 1944 major clashes broke out in Athens between the Communist Party and Government forces, which were critically aided by British troops.[5] The conflict lasted for a month and ended with the defeat of the communists, who conceded to surrender arms all over the country under the so called "Varkiza Treaty", signed in February 1945. At exactly the same time, the Yalta Conference put Greece firmly into the British sphere of influence, and the Soviet Union subsequently refrained from endorsing the Communist Party's bid to seize power by force.

2.1.3 The Onset of the Civil War

General elections were held in March 1946, but were boycotted by the Communist Party as allegedly rigged and illegitimate. The King returned in September 1946 after a referendum, conducted with the full participation of the Communist Party. A month later the Communist Party decided to organize the Democratic Army of Greece (DAG), and openly called for an armed struggle to overthrow the 'monarch-fascist' Government.

As the Civil War started to unfold, Britain withdrew from Greece, and the move gave the impression that the guerrillas could win the conflict. But the power vacuum was soon filled by the US, after President Truman proclaimed in 1947 his doctrine of containing the expansion of communism. As a result, substantial military aid was sent to the Greek National Army (GNA), and a little later the planning of war operations was effectively undertaken by the US military mission to Greece.

The conflict ended only after a decisive defeat of the guerrilla forces in August 1949. It left behind more than 43,000 dead and many more wounded, imprisoned or

[4]For details see Christodoulakis (2014).

[5]For an account of the clashes, see Woodhouse (1976). On the strategy of KKE and its consequences, see Iatrides (1972).

exiled in detention camps. Tens of thousands of guerrillas and their families fled to the Soviet Union and other Eastern European countries to escape further persecution.

2.1.4 *Prosperity and Democracy Curtailed*

In spite of the destitution inherited from the Occupation, there was no sign that ordinary Greek people were keen to engage in yet another war: the common expectation—shared by the Government as well as the Communist Party—was that Greece would be assisted by the Allies to rebuild its economy, improve the functioning of the state and restore safety and calm.[6] After liberation, the authorities managed to stabilize inflation, issued a new currency and tried to kick-start reconstruction.[7] In 1946 the economy rebounded and, for a while, it was thought that Greece would embark on a rapid recovery path, as was happening elsewhere in Europe.

But despite the early success, the process soon was hampered by the continuing clashes between the Left and the Right and the looming threat of a protracted conflict. The reconstruction process was impaired, as several productive units had been destroyed and new investment was thwarted. Moreover, investment in new social and productive infrastructure was crowded out by excessive military spending, further fuelling discontent and leading to repeated failures of reconstruction efforts and stabilization policies.[8]

Hence, the victory dividend that was so dearly won by Greece during the Second World War was squandered during peacetime. During the Civil War, the country was marginalized in the post-war process of European reconstruction, and domestically the winning parties embarked on a new wave of exclusion politics against left-leaning citizens. Parliamentary democracy remained weak and fragmented, and a military dictatorship was imposed in 1967. A full-fledged democracy was finally restored in 1974, while the Monarchy was permanently abolished at the same year.

2.2 Data Analysis

Time-series of battle-deaths and total battle-casualties of DAG and GNA cover the period from January 1946 until December 1949 on a monthly frequency and are used to examine the statistical properties of the conflict.[9] Other data regarding the

[6]KKE (1964, p. 524)—adopting a wording that would soon be abandoned—emphasized the importance of "ensuring order and calm so as to rebuild the country".

[7]Makinen (1984, 1986).

[8]A description is given by Thomadakis (1995).

[9]Battle deaths and casualties include all people, combatant and civilian, that are killed or injured in armed engagements.

size of the armies, the recruitment process, or the number of captives and deserters are only partially available and used to describe specific developments.

2.2.1 The Statistics of Conflict

The evolvement of the Greek Civil War is shown in Fig. 2.1. It is clear that during the first half of 1946 casualties remained too low for the conflict to qualify as a civil war. It is only in the second half of 1946 that hostilities become widespread and organized conflict ensues. To face widespread persecution by right-wing militias, the Communist Party started to organize 'self-defence' groups throughout the country. The Government retaliated by setting up emergency martial courts against left-wing activists, who were massively prosecuted for acts against "public order and safety."[10]

Henceforth, it is assumed that the Civil War started in July 1946 and ended with the final offensive of GNA against the guerrillas in the last days of August 1949. In September 1949 the official surrender of DAG was signed and the Civil War was terminated, although some sporadic hostilities continued for a while.

Some key conflict statistics are summarized in Table 2.1. Reflecting the escalation of conflict, both series of battle-deaths and casualties are found to be non-stationary by a unit root test. Losses seem to have risen sharply in the beginning of 1947, when DAG forces attacked several towns and GNA launched the first wave of military operations to clear guerrilla strongholds in the mountains.[11] By the end of 1947, the conflict intensified and acquired new operational and political dimensions.

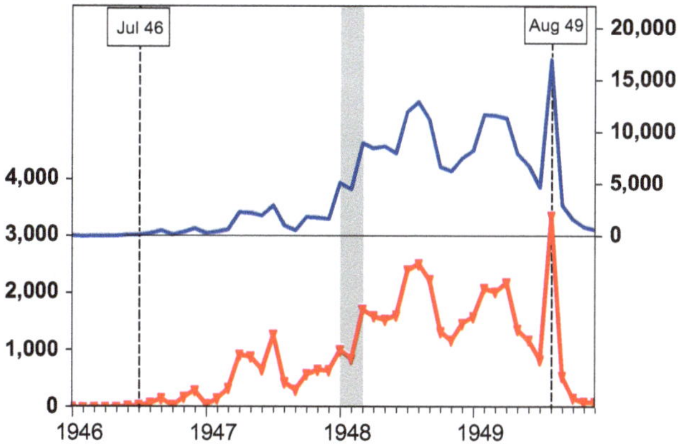

Fig. 2.1 Total battle-casualties and battle-deaths. *Notes*: (*Upper graph*, Rhs) and (*marked graph*, Lhs) respectively. Period 1946:01–1949:12. The beginning of Civil War is marked in July 1946 and its end in August 1949. The *shaded area* in 1948 marks a structural break into two phases. *Source*: Data are described in Appendix and listed in Table 11.1

[10]The Third Decree of the State was issued in July 1946 and initially established 11 martial courts in key cities. A few months later the number rose to 30, covering most of the country.

[11]For an account of town sieges by the guerrilla army, see Marantzidis (2010, p. 192).

Table 2.1 Periodicity tests

	Total battle deaths	Total casualties
Unit root test 46:07-49:09	−1.69 (p = 0.42)	1.33 (p = 0.60)
Jarque_Bera normality test	2.78 (p = 0.24)	3.16 (p = 0.20)
Period	*Breaking points* (statistics for detrended series)	
47:12	F = 0.31 LLR = 0.278	0.0722; 0.0530
48:01	0.126; 0.099	0.0072; 0.0041
48:02	0.0775; 0.0574	0.0076; 0.0043
48:03	0.0187; 0.0117	0.0019; 0.0009

Table 2.2 Key statistics of battle deaths and casualties

	Total deaths		Total casualties	
Statistics	Phase I 46:07–47:12	Phase II 48:01–49:09	Phase I 46:07–47:12	Phase II 48:01–49:09
Mean	454	1659	1130	8737
Std dev	340	663	885	3340
Volatility (%)	75	40	78	38
Pareto index (k)	0.74 (0.214)	0.98 (0.255)	0.62 (0.182)	1.01 (0.262)
x_{min}	89 (7.83)	551 (29.64)	161 (17.21)	3008 (156.6)

Note: Standard errors in brackets. Data as in Appendix

2.2.2 *Periodicity*

This calls for the proper investigation of separate phases in the conduct of civil war. Letting a simple time-trend to remove non-stationarity and applying Chow tests, a structural break is detected at the beginning of 1948, indicating a periodicity of two phases:[12] The first phase lasted from July 1946 to December 1947 and the second from January 1948 to September 1949 when the official surrender was concluded.

The data series for both battle deaths and casualties are now found to be stationary over the two sub-periods as shown in the first row of Table 2.2. Comparing battle statistics in the two phases, a crucial change concerning the extent and nature of the conflict is revealed: the monthly battle-death average increases nearly fourfold, while that of total casualties rises nearly eightfold. In Phase II volatility—expressed as the ratio of standard deviation to the mean—drops by nearly half its size in Phase I. This is suggesting that the earlier pattern of widespread skirmishes gave way to larger-scale confrontations.

[12]The hypothesis of no breakpoint between January and March 1948 is rejected at a range of levels from 1 % to 10 % for total casualties and battle-deaths as shown in Table 2.1.

2.2.3 The Conflict Distribution

A new strand in the quantitave literature of conflict claims that the duration of a conflict can be assessed by computing the complementary cumulative distribution function (*ccdf*) of battle casualties.[13] If found to follow a power-law distribution, this implies that the probability of losses exceeding a high level—and thus leading to the termination of conflict—is not very low, in contrast to the predictions of a normal distribution. The most common pattern is the Pareto distribution, in which the probability of casualties *(X)* exceeding a certain level *(x)* is given as:

$$ccdf(x) = prob(X \geq x) = \left(x/x_{min}\right)^{-k} \tag{2.1}$$

where (k) is the conflict index and (x_{min}) represents a lower bound of fatalities, i.e. $prob(X > x_{min}) = 1$. The lower the index (k), the more likely a large number of casualties will occur and, therefore, the conflict is likely to continue and reach a higher level of casualties. Using a large cross-section of fatality data from conflicts worldwide, Clauset et al. (2007) show that they follow a Pareto distribution with an index equal to $k = 2.50$, which is taken to express a global constant of conflict. However impressive such a finding might be, its usefulness should not be overstated. As noted by Lichbach (1992), such an a theoretical approach amounts to claiming that conflicts occur randomly, in sharp contrast with the schools of thought that interpret them as outcomes of rational calculation or as a response to grievances.[14]With this caveat in mind, we examine whether power-law models are relevant for understanding the dynamics of the Greek conflict. Time-series of total battle-deaths and casualties are tested against a Pareto distribution by applying Maximum Likelihood techniques, and the results are shown in Table 2.2. Despite the relatively small number of observations, all estimates are statistically significant.

In Phase I the conflict index is found to be equal to 0.74 and 0.62 for deaths and casualties respectively, much lower than the presumed global constant. These figures suggest that the probability of casualties exceeding a certain level is high or, in other words, that the conflict is likely to escalate even further. In Phase II the conflict index is found to be equal to 0.98 and 1.01 for deaths and casualties respectively, closer to the presumed global constant. These figures suggest that the probability of casualties exceeding a certain level is high or, in other words, that the conflict was further escalating. The estimates are, however, found to be sensitive to the choice of the estimation period. For example by cutting off the last month so that the sample period is 1948:01–1949:08, the index is estimated at 1.55 and

[13]This is frequently called the 'survivor' function, but here the term is unsuitable for describing battle-deaths.

[14]Criticism is also addressed to a further claim that relates power-law distributions with other confrontational phenomena, from ecology to finance and social dynamics; for example, see Bohorquez et al. (2009).

1.60 for deaths and casualties respectively, very close to the estimates of 1.70 found by Bohorquez et al. (2009) for the US and the Spanish civil wars. Due to the lack of robustness, the analysis is not pursued any further.

2.3 The Escalation of Conflict

The escalation of conflict in early 1948 was preceded by extensive operationalization and the enlargement of both armies, as shown in Fig. 2.2. The build-up process in each army is described below.

2.3.1 *The State Army (GNA)*

The state army was steadily growing and by the end of 1947 had reached 120,000 men, an increase by more than 30 % from a force of 92,000 men in the beginning of the year. But despite its expansion, the state army proved incapable of swiftly containing the guerrilla forces. The numerical supremacy of GNA was compromised in practice, as many of its forces were allocated in non-combatant duties, while others were untrained and/or inadequately equipped for fighting on a mountainous terrain.[15] It was only after 1947 that such drawbacks were decisively corrected, mainly thanks to the US mission's providing the state army with modern arms and training.

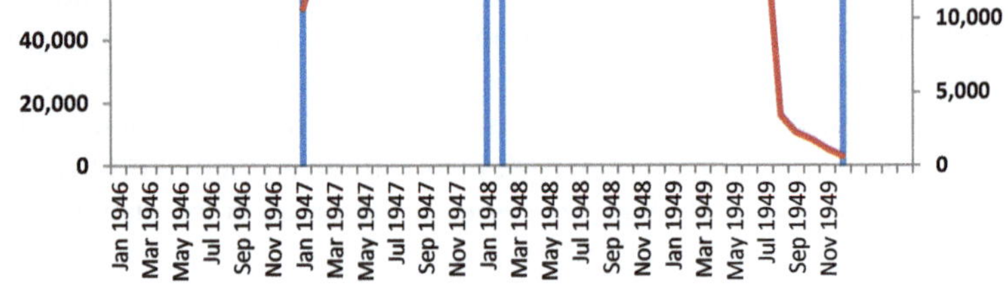

Fig. 2.2 The size of GNA and DAG armies. *Notes*: Lhs and Rhs respectively. January 1947–December 1949. *Source*: Data described in Appendix and Table 11.1

[15]See Marantzidis (2010, p. 92), Averof-Tositsas (2010, p. 208) and Gregoriadis (2011, p. 166).

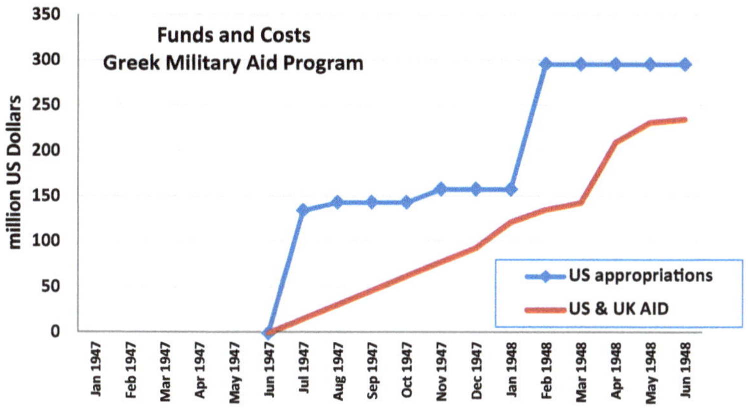

Fig. 2.3 The costs of military aid delivered to Greece for GNA. *Source*: JUSMAGG 1949. Diagram funds and costs, greek military aid program, ground and air

In 1948, the GNA went further up to 132,000 men, while military shipments from the US were multiplied; see Fig. 2.3. Combat drills became more demanding, and non-combatant duties were delegated to the National Guard Battalions. At the same time, the political cleansing of the army was intensified: left-leaning soldiers were massively transferred to isolated islands, and hundreds of officers were court-martialled for alleged communist infiltration.[16]

Finally, a high US command arrived in Athens in February 1948 to directly coordinate military operations. This radically altered the course of the civil war, both operationally and with regards to its geopolitical repercussions on the ensuing Cold War.

2.3.2 *The Guerrilla Army (DAG)*

Following a parallel—though much less resourceful—process the guerrilla army was trying to expand its force and improve logistics. Neighbouring Balkan states were offering military training and backyard facilities to DAG fighters. In an effort to counter the improved capabilities of GNA, substantial military equipment was

[16]The most notorious concentration camp was on the Macronissos Island where 28,800 soldiers and officers were kept during 1947–1950. Though the Government hailed the camp as a "moral transforming institution", several of the interns perished out of torture and starvation; for an account, see Kaltsogia-Tournaviti (2001).

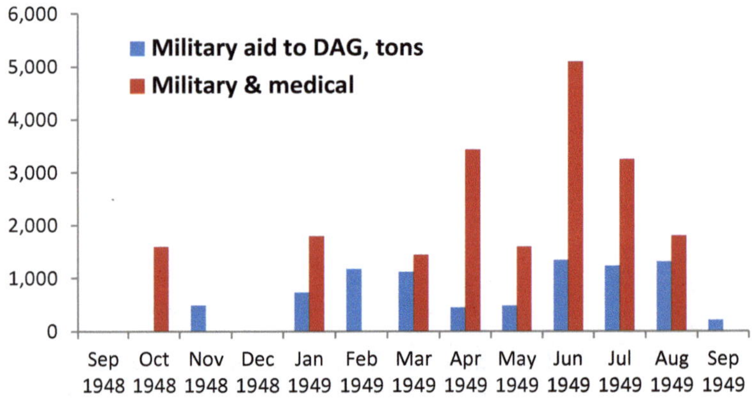

Fig. 2.4 Military and non-military aid delivered to DAG. *Source*: Marantzidis (2010)

shipped to DAG from Poland and other Eastern European countries in 1948–1949, as displayed in Fig. 2.4.[17] At the same time, DAG was extensively restructured to cover the entire mainland as well as several islands.[18]

In December 1947 the guerrillas formed a 'Provisional Government' and launched their first tactical warfare operation to proclaim Konitsa—a town near the northern borders—as the capital of 'liberated' territories. After 2 weeks of intense fighting the offensive was defeated, and subsequently the strategy of DAG concentrated on the war in the countryside.[19]

2.3.3 Population Displacement

To check the size of DAG from rising any further in the countryside, more than 350,000 villagers in Northern Greece were displaced from their lands and transferred to refugee camps in nearby cities during 1948 and 1949. The Government touted the operation as protecting "bandit-stricken" villagers from plunder, though KKE claimed that it was a 'scorched earth' policy designed to undercut provisions, recruitments and the intelligence system of DAG; [20] for a detailed account, see Laiou (1987, p. 61) and Averof-Tositsas (2010, p. 237). Displacements severely undercut DAG recruitments and this may be one of the reasons that its forces could not increase any further after 1948 as shown in Fig. 2.2.

---

[17]Details are given in Marantzidis (2010, pp. 48–49).

[18]Details of the new DAG structure are given by Kyritsis (2006, p. 28).

[19]Rizospastis (2011). That was the first open disagreement about strategy, with the DAG leader supporting partisan fighting and the Secretary General of KKE insisting on urban struggle.

[20]Rizospastis (2011).

Those developments further fuelled hostilities and undermined the chances of a negotiated end to the conflict. For example, the evacuees in the rural areas developed a strong opposition against guerrillas, and several of them volunteered to join the Government forces. On the other hand, the wave of prosecutions created a potential pool for guerrilla recruits, as several would-be suspects opted for joining DAG in the mountains rather than facing court-martial. In the beginning of 1948 it was clear that both the Government and the guerrillas were moving toward a prolonged and deadly confrontation.

References

Averof-Tositsas E (2010) By fire and axe: Greece 1946–1949 and the precedents. Estia Editions, Athens (in Greek)

Bank of Greece (1947) Director's report for the years 1941–1946. Bank of Greece Publication, Athens (in Greek)

Bohorquez JC, Gourley S, Dixon AR, Spagat M, Johnson NF (2009) Common ecology quantifies human insurgency. Nature 462:911–914. doi:10.1038/nature08631

Christodoulakis N (2013) Currency crisis and collapse in interwar Greece: predicament or policy failure? Eur Rev Econ Hist 17(3):272–293

Christodoulakis N (2014) Germany's war debt to Greece: a burden unsettled. MacMillan Editions, London

Clauset A, Young M, Gleditsch K (2007) On the frequency of severe terrorist events. J Confl Resolut 51(1):58–88

Close DH (1986) The police in the fourth-of-august regime. J Hellenic Diaspora 13(1–2):91–105

Doxiadis J (1946) The sacrifices of Greece in the Second World War. Reprinted by Kathimerini Editions (2014), Athens

Gregoriadis S (2011) N. Zachariadis: the fatal leader, vol IV. Kyriakatiki Editions, Athens (in Greek)

Iatrides JO (1972) Revolt in Athens: the Greek communist second round, 1944–1945. Princeton University Press, Princeton

Kaltsogia-Tournaviti N (2001) The institutional aspects of the civil war. In: Koutsoukis K, Sakkas I (eds) Aspects of civil war. Filistor Editions, Athens, pp 1946–1949 (in Greek)

KKE (1964) Forty years 1918–1958: collection of documents. Party Edition, Athens (in Greek)

Kyritsis N (2006) Democratic army of Greece: key phases of the struggle. Synchroni Epohi, Athens (in Greek)

Laiou A (1987) Population movements in the Greek countryside during the Greek Civil War. In: Baerentzen L, Iatrides JO, Smith OL (eds) Studies in the history of the Greek civil war. Museum Tusculanum Press, Copenhagen, pp 1945–1949

Lichbach M (1992) Nobody cites nobody else: mathematical models of domestic political conflict. Def Econ 3:341–357

Makinen G (1984) The Greek stabilization of 1944–1946. Am Econ Rev 74(5):1067–1074

Makinen G (1986) The Greek hyperinflation and stabilization of 1943–1946. J Econ Hist XLVI(3): 795–805

Marantzidis N (2010) The democratic army of Greece (1946–1949). Alexandria Editions, Athens

Mazower M (1994) Inside Hitler's Greece: the experience of occupation, 1941–1944. Alexandria Editions, Athens

Pelt M (2014) Stages in the development of the 'Fourth of August' regime in Greece. In: Pinto AC, Kallis A (eds) Rethinking fascism and dictatorship in Europe. Palgrave MacMillan, London, pp 198–218

Rizospastis (2011) The three-year epic of the democratic army of Greece 1946–1949. Synchroni Epohi Editions, Athens (in Greek)
Thomadakis S (1995) Stabilization, development and government economic authority. In: Iatrides JO, Wrigley L (eds) Greece at the cross-roads 1944–1950: essays on the civil war and its legacy. Penn State Press, Pennsylvania, pp 180–186
Woodhouse C (1976) The struggle for Greece: 1941–1949. Hurst, London

Chapter 3
Static Models of Conflict

Two main approaches are used in civil war literature to describe the reasons behind the occurrence of, and participation in, a conflict: The first is based on the rational theory of actions that evaluates the trade-off between opportunities and costs, described in the form of expected versus foregone benefits. The other approach argues that a conflict is caused by grievances against vested interests that hold society in a backward state and prevent opportunities for the many. This chapter describes the static opportunity-cost model and employs it to explain some stylized facts observed in civil wars.

Using a regional database of Greece, a number of socio-economic indicators are tested as to whether they are correlated with political grievances prevailing at the time of the civil war.

3.1 Conflict Motivation

In the academic literature on civil wars two alternative theories compete for explaining their occurrence, participation rate, and severity: the opportunity-cost model and the theory of accumulated grievances. Despite the arguments for and against each one, the two approaches have, in fact, more similarities than differences when describing a process of organized conflict, for two reasons:

First, because a conflict is characterized as a civil war only if it is extensive, recurrent and identified with a central cause. As opposed to uncoordinated actions of atomized violence, a civil war process requires a disciplined subordination to an organized set of tactics and strategies, in order to achieve an outcome against another collective entity. The subordination mechanism may be voluntary or coercive, material or stemming from a value system such as ideology, religion or ethnicity. A particular person may or may not decide to participate in a rebellion by weighing individual opportunities and costs, but the collective character of a

N. Christodoulakis, *An Economic Analysis of Conflicts*,
DOI 10.1007/978-3-319-32261-2_3

conflict goes beyond a mere aggregation of individual actions, even though such practices may be occasionally witnessed in a civil war.

Second, because the cause of a conflict is usually justified by denouncing some existing grievances that strip the opportunities of the people at large. The underprivileged are invited (or forced) to participate and thus improve their own prospects.

In this sense, organized conflicts show common characteristics with situations in which competitive parties contest for a reward. Such models have been widely used in economics and are useful in examining the behaviour of two sides in a conflict. Before presenting the framework for analysing civil wars, the classic contest model is briefly examined.

3.2 The Static Tullock Model

In a seminal paper, Tullock (1967) described a contest as a competition among a number of parties, with each of them devoting a certain amount of effort in pursuing a rewarding outcome net of the costs incurring. The main characteristics are described below.

3.2.1 Payoffs and Win Probabilities

A conflict can be considered as a contest between two rival parties ($j = 1, 2$), which have pay-off functions defined as:

$$\Pi_j = p_j V_j - C_j(x_j) \tag{3.1}$$

In the above expression, V_j is the reward that party j is to receive in case of victory, x_j is the effort employed by the contestant, and $C_j(.)$ a cost function increasing in effort. The probability of party j winning the contest is denoted by p_j, and called the "conflict success function" (CSF). Naturally, the winning probability depends on the effort exercised by each side: it increases with own effort and is reduced by the other party's effort. In general, it is described by

$$p_j = p(x_j, x_{i \neq j}) \tag{3.2}$$

The following properties are assumed throughout

$$\frac{\partial p_j}{\partial x_j} > 0, \frac{\partial p_j}{\partial x_{i \neq j}} < 0 \tag{3.3}$$

If winning probabilities are such that $p_1 + p_2 = 1$ (i.e. $p_2 = 1 - p_1$), there is only one winner who takes the prize. The odds of the first party winning are given by the ratio (p_1/p_2), which is called the "'win ratio'." In general, however, winning probabilities need not add up to unity and, in such a case, no clear winner may emerge. The remainder $(p_0 = 1 - p_1 - p_2)$ expresses the probability of a stalemate.

Let's first consider the clear-winner case. For $i \neq j$, a suitable form that ensures such an outcome is

$$p_j = \frac{f_j(x_j)}{f_j(x_j) + f_i(x_i)} \tag{3.4}$$

where function $f_j(x_j)$ denotes the effectiveness of effort x_j exercised by party $j = 1$, 2. Various forms are proposed in contest literature to model effectiveness. A popular function is the "power-form" $f_j(x_j) = (x_j)^m$ as proposed by Tullock (1967). In this case the win ratio (p_1/p_2) is easily expressed as a function of the ratio of the two parties' efforts $(x_1/x_2)^m$, and for this reason this formulation is also called the "ratio form." Exponent $m > 0$ denotes how much the effectiveness is improved by a rise in effort, and represents the "discriminatory power" of each side, as discussed by Grossmann et al. (2011).

The snag with this structure is that if one party chooses not to fight (i.e. $x_1 = 0$), its winning probability vanishes and there is no award in the clear-winner outcome. This does not cover the option for a party to appease the aggressor in the hope that it will be allowed to share some part of the reward or keep a fraction of the assets it already possesses. To deal with this problem, Hirshleifer (1989) proposed a logistic function capturing the possibility of one party abstaining from conflict but still expecting that some reward will accrue to it, i.e. $p_1(0, x_2) > 0$ and similarly for p_2. In this case, winning probabilities are described by the functions:

$$p_1 = \frac{e^{mx_1}}{e^{mx_1} + e^{mx_2}}, p_2 = \frac{e^{mx_2}}{e^{mx_1} + e^{mx_2}} \tag{3.5}$$

Another formulation can be adopted for the possibility of a stalemate. In describing a conflict, Konrad and Skarpedas (1998) consider a general function form of the kind:

$$p_1 = \frac{f_1(x_1)}{f_1(x_1) + f_2(x_2) + f_0}, p_2 = \frac{f_2(x_2)}{f_1(x_1) + f_2(x_2) + f_0} \tag{3.6}$$

where $f_j(x_j)$ is the "effectiveness function" as before, and f_0 is a constant. With $f_0 > 0$ there might be a stalemate with probability

$$p_0 = 1 - p_1 - p_2 = \frac{f_0}{f_1(x_1) + f_2(x_2) + f_0} > 0 \tag{3.7}$$

3.2.2 *Optimal Effort*

Each party's effort is determined by maximizing (3.1). This is a Nash-Cournot game with two players and, by setting the first-order derivatives equal to zero, we obtain

$$\frac{\partial \Pi_j}{\partial x_j} = 0 \Rightarrow \frac{\partial p_j(x_j, x_{i \neq j})}{\partial x_j} \cdot V_j = \frac{\partial C_j}{\partial x_j} \tag{3.8a}$$

The above expressions imply that the marginal expected rewards equal marginal costs. By solving the above equations, the effort exercised by one side is expressed as a reaction function to the effort exercised by the other. In general, it takes the form:

$$x_j = F_j(V_j, x_{i \neq j}) \tag{3.8b}$$

The effort exercised by each party is thus a function of its own reward and the other's effort. Obviously, it must be increasing in both arguments. For the effort to lead to a maximum payoff, the second-order conditions must be satisfied for $j = 1,2$:

$$\frac{\partial^2 \Pi_j}{\partial x_j^2} < 0 \Rightarrow \frac{\partial^2 p_j}{\partial x_j^2} V_j < \frac{\partial^2 C_j}{\partial x_j^2} \tag{3.9}$$

3.3 The Collier-Hoeffler Framework

In an influential paper on civil wars, Collier and Hoeffler (2004) employ the above framework in order to derive the size of the armies involved in a conflict between guerrillas and government forces. Rewards are expressed as claims on the country's output but, in general, they may include other interests, such as territories, political posts, rents, etc. The government holds power and already possesses an asset that it tries to safeguard from expropriation by the insurgents.

A restrictive assumption is made for the efforts expended by the two sides. The decision on whether to enter the conflict and how much to fight in it, is taken with regards to the size of the army that is mobilized in the conflict. It is also assumed that $m = 1$, thus fighting effectiveness is linear in army size and common to both sides. The notation and terminology below is adapted to a civil war context and the model is described as follows:

3.3.1 Guerrillas

In the prospect of entering a civil war, guerrillas organize an army of size (r), and are likely to win the conflict with probability (p_1), and thus obtain control of a reward V_1. This may stand for the country's output and/or the privileges of holding power. By analogy, they are likely to be defeated with probability $(1-p_1)$, whereat they receive nothing. The payoff function is given as:

$$\Pi_1 = p_1 V_1 - w_1 r \tag{3.10a}$$

For each individual joining the rebellion, variable w_1 denotes per capita income forgone. In a collective setting, it may denote the per capita cost of maintaining the guerrilla army.

3.3.2 Government

The government maintains a state army of size (q) and controls an asset of value V_2. If challenged by a civil war, the government defends the *status quo* and is likely to win with probability (p_2) and lose with probability $(1-p_2)$. Thus the government has a payoff function equal to

$$\Pi_2 = p_2 V_2 - w_2 q \tag{3.10b}$$

Fighting effectiveness is assumed to be the same across the two armies, but the government has an embedded advantage (say, q_0). The conflict success functions are:

$$p_1 = \frac{r}{r + q + q_0}, \; p_2 = \frac{q + q_0}{r + q + q_0} \tag{3.11}$$

Finally, Collier and Hoeffler (2004) introduce a barrier to entering a civil war, by adopting an ad hoc constraint that the army of guerrillas should exceed a certain proportion (or multiple) of government forces. This is called the "viability constraint" and is given by

$$r > \mu \cdot (q + q_0), \; \mu > 0 \tag{3.12}$$

The above condition echoes various practical rules on the appropriate size of an army. For example, the so called "3:1 rule" postulates that for an aggressor to overcome the defenders, it must have three times as much power, that is $\mu = 3$. In practice, however, it seems that it takes substantially lower values, since several civil wars break out with guerrilla forces much inferior to the state armies.

3.4 Optimal Armies

The first-order condition for maximizing (Π_1) yields the optimal army size $(\widehat{r})$ that has to be mobilized by the guerrillas. Since they may not know the exact size of their opponents, the reaction function is obtained as follows:

$$\widehat{r} = -(q^e + \mu) + \left[\frac{V_1}{w_1} \cdot (q^e + q_0)\right]^{1/2} \tag{3.13a}$$

where (q^e) is a guess of the government's forces actual size. By analogy, the optimal size of government forces is determined as:

$$\widehat{q} = -(r^e + q_0) + \left[\frac{V_2}{w_2} \cdot r^e\right]^{1/2} \tag{3.13b}$$

As shown in Fig. 3.1, both army functions are concave. The guessing on each other's army size is adjusted on the basis of new information, the system eventually converges to the steady-states. Setting $r^e \rightarrow \widehat{r}$ and $q^e \rightarrow \widehat{q}$, rearranging (3.13a, 3.13b) and dividing by parts, one can finally obtain that the optimal sizes of the two armies should, in equilibrium, satisfy the condition for the win ratio:

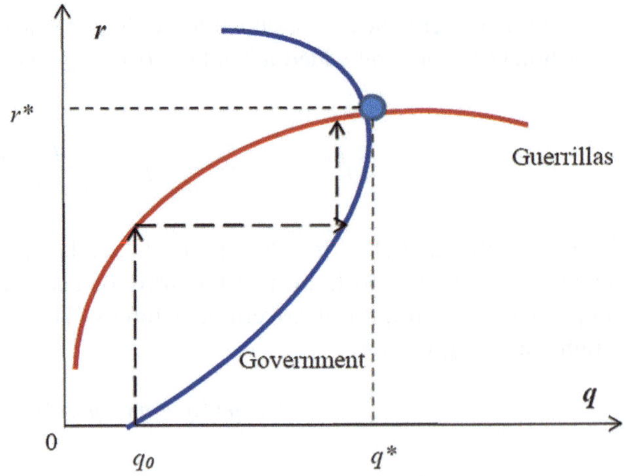

Fig. 3.1 Army formations for the guerrilla and the government armies

$$\frac{p_1}{p_2} = \frac{\widehat{r}}{\widehat{q} + q_0} = \frac{w_2}{w_1} \cdot \frac{V_1}{V_2} \tag{3.14a}$$

The above army ratio is similar to the linear version of Lanchester's combat models, to be discussed in Chap. 5. The viability constraint (3.12) is translated into the condition:

$$\frac{V_1}{V_2} > \mu \frac{w_1}{w_2} \tag{3.14b}$$

This can be interpreted as a requirement that the relative reward of a successful insurgency should surpass the current income inferiority by at least the fighting viability factor (μ).

3.5 Stylized Facts

Although it is derived through several simplifying assumptions, the Collier-Hoeffler model is capable of explaining a number of stylized facts as examined below:

3.5.1 Inequality and Class Struggle

The ratio of the wage rates (w_2/w_1) that appears in (3.14a) can be taken as a measure of income inequality between government personnel and the guerrillas. Alternatively, it may be interpreted as an indication of the class division that characterizes the particular economy. The higher it is, the stronger the motivation an excluded individual has to join the guerrilla army. To the extent that the above characteristics are related with certain measurable factors, one can use statistical data to test the hypothesis. For example:

Income Inequality High inequality in per capita incomes between the deprived parts of the population and the governing elites is likely to exert an upward effect in the ratio (w_2/w_1). As the ratio increases, the win ratio in (3.14a) rises in favour of the guerrillas, thus a civil war is more likely to occur. The channels through which inequality leads to political instability and conflict are discussed and empirically examined by Alesina and Perotti (1996). Also, inequality is expected to be higher in countries with extensive ethnic and social fractionalization. Hence, these two factors are likely to be positively correlated with the occurrence and severity of civil conflicts.

Per Capita Output and Growth The state-controlled endowment V_2 is usually a function of country's output in per capita terms. If the country is poor and/or output is falling, the value of the endowment is reduced and the win ratio for the guerrillas is increased. Thus, civil war is more likely to erupt in poor or suddenly impoverished societies. For the same reason, a country that has recently experienced a war and fallen in poverty is more likely to get trapped into another conflict.

Extensive empirical confirmation of the above analysis is provided by several authors. Perhaps the most representative evidence presented for modern civil wars is the collection of research articles included in a two-volume publication supervised by Collier and Sambanis (2005). Their findings suggest that variables such as per capita income, the growth rate of the economy, and male secondary education enrollment, have statistically significant and substantial effects on influencing the opportunity of rebellion. Combined with military advantages and financial availability, opportunities may turn to actual design and implementation of fighting, though the authors keep a distance from suggesting one-way causation.

3.6 Criticism to the Collier-Hoeffler Model

A number of limitations are inherent in the Collier-Hoeffler model and are discussed below.

A key feature is that the cost incurred by each side is only caused by income foregone by entering the conflict. The implication of such a cost function is that the aggressor, in case of failing to win, suffers no more for taking part in the insurgence. This is in stark contrast with civil war realities, by omitting the examination of the consequences that persecutions may have on the occurrence and outcome of a conflict. In real world situations, a failed rebellion is most likely followed by reprisals against the defeated, and this may become an important deterrent for a party entering a conflict in the first place.

The model also describes a static situation where the two sides decide to enter the conflict for a single and terminally decisive fight. This hardly reflects reality, where civil wars are usually protracted operations and may last for a long period of time before a clear winner can emerge. In other cases, there might be a "conflict trap" where the sides are involved in repeated hostilities without the probabilities of winning being systematically altered in favour of either. As a consequence of the lack of dynamics the role of randomness is ignored and insights can be discussed only by comparative statics. Thus any stochastic characteristics that a conflict process may have—for example those examined in Chap. 2—cannot be studied. This can be somewhat remedied only by assuming that the static model is always in equilibrium and then postulate some ad hoc adjustment process towards it. But, again, this is an extreme assumption to be made.

Another criticism is that the Collier-Hoeffler model deals only with the size of the army that is mobilized by each adversary, without paying any attention to the number of casualties and the effect they have on the subsequent balance of power.

Finally, there are also problems related with the way that the stylized facts are estimated across countries. A study published by Mankiw et al. (1995) expresses some important warnings on the use of cross-country regressions to identify causal effects. First is the problem of simultaneity between explanatory variables and the phenomenon they are used to explain. In the civil war context, factors such as low growth may not only cause civil war, but may also be aggrieved by it.

Second is the problem of collinearity, since some of the explanatory variables, rather than being independent, may in fact influence each other. For example, low growth and low school enrollment, which are frequently used to explain the incidence of civil wars, are more likely to feed each other rather than occur independently. Finally there might be a degrees-of-freedom problem if too many explanatory variables are used to explain civil wars in a limited number of countries.

3.7 APPLICATION: Greek Grievances and Conflict

3.7.1 *Social Grievances*

To examine whether political rivalry in post-WW2 Greece was influenced by socioeconomic factors, a spatial analysis is conducted across the 38 prefectures which comprised the country at that time. A measure of political discontent is formed by taking a weighted average of the rate of abstention from the 1946 elections (variable ABS46) and the republican vote in the 1946 referendum (variable REP46). Both options were strongly advocated by the Communist Party as a manifestation of political opposition to the government. Note, however, that the republican vote was not fully reflecting the influence of the Left as it was also advocated by the Liberal Party. Nevertheless, in the absence of more details on the composition of the republican vote, the measure of political discontent is set as the simple average of the two votes[1]:

$$Discontent = 0.50{*}ABS46 + 0.50{*}REP46 \tag{3.15}$$

The above measure is then regressed against various indices of grievance to examine whether and how discontent is connected to socioeconomic conditions. Grievances may stem from social exclusion, ethnic polarization and inequality, lack of basic goods and services, or simply when the prospects of improving such

[1]Other combinations of weighting coefficients do not significantly affect the results, and the findings remain qualitatively the same.

inherited conditions look bleak. Three types of economic grievance are considered below as potential sources of political animosity, depending on the causes of adversity stemming from one of the following:

1. Current deprivation
2. Class structure and exploitation
3. Old-entrenched grievances and rivalries

3.7.2 Contemporaneous Deprivation

Adverse social conditions at the time are described by the unemployment rate, the crime rate and the flux of emigrants in search of better opportunities abroad, while they were further exacerbated by the spreading of diseases such as malaria and tuberculosis.[2] Quality of life is measured by the consumption of electricity, the use of telecommunications, and the provision of medical care expressed by the number of hospital beds in proportion to local population. Additionally, the expansion of media and access to secondary education indicate openness of local society and better prospects for the new generation respectively. A high degree of adversity and poor quality of life could possibly lead to social dissatisfaction and political protest.

Correlations are shown in Table 3.1. It is perhaps striking that none of the adverse or quality indicators is found to be significantly correlated with political discontent; the only exception being the number of hospital beds per head of population but is wrongly signed nonetheless.[3] If a similar analysis is conducted by using the share of agricultural to total population in each region, all positive social aspects are found to be strongly and negatively correlated; see the last column in Table 3.1. Of the adverse indicators, malaria is strongly and positively correlated with the index of rural population composition. Unemployment and emigration are negatively correlated, probably implying that family ties were stronger in the countryside, by offering some form of land cultivation and, thus, discouraging emigration. Only tuberculosis is found to be negatively correlated with the countryside. Even this finding, however, may not necessarily reflect a better environment but is probably due to the fact that the affected patients were registered in hospitals mostly located in urban centers.

Taken together, the above findings reveal that most living conditions had markedly deteriorated in the countryside, yet the political protest was disconnected from contemporaneous social ills. Thus a deeper division was perhaps tearing Greek society apart, and further explanations are attempted below.

[2]Often the spreading of diseases is strongly correlated with low per capita income and impoverishment. For the case of malaria, see the study of Datta and Reimer (2013). Similarly, Spence et al. (1993) establish that tuberculosis is strongly associated with poverty.

[3]The explanation might be that hospitals were concentrated in larger cities where discontent was stronger or perhaps less suppressed.

Table 3.1 Sociopolitical correlations

Social indicators		Correlation coefficient (t-stat in brackets)	
	Indicator	Political discontent	Rural index
Adverse aspects			
Unemployment	% active population	0.24 (1.50)	−0.80*** (8.07)
Emigration	Emigrants % population	0.06 (0.37)	−0.33** (2.08)
Crime rate	Convictions % population	−0.17 (1.04)	−0.10 (0.62)
Tuberculosis	Deaths per 10,000	+0.14 (0.87)	−0.62*** (4.75)
Malaria	Deaths per 10,000	0.113 (0.67)	+0.33** (2.11)
Quality of life			
Electricity	Kwh per head	−0.03 (0.18)	−0.73*** (6.33)
Communication	Telegrams per head	−0.10 (0.60)	−0.81*** (8.22)
Health services	Hospital beds per 10,000	+0.32** (2.06)	−0.64*** (5.04)
Media	Newspapers per 10,000	0.12 (0.73)	−0.74*** (6.63)
Secondary education (Enrollment)	Pupils per 10,000	−0.03 (0.21)	−0.45*** (3.06)
Secondary education (teaching staff)	Teachers per 10,000	−0.08 (0.49)	−0.42*** (2.81)

Notes: Political discontent is measured by the simple average abstention rates in the 1946 elections (ABS46) and the republican vote in the 1946 Referendum (REP46). The rural index is the proportion of active agricultural population (NAGR) to total (NACT). One, two or three stars indicate p-values at 10 %, 5 % or 1 % respectively. Data cover 38 regions. Definitions and sources as in Data-Appendix

3.7.3 Discontent and Class Structure

The view adopted as the official line by the Communist Party is tested first. According to the dominant Marxist interpretation of the time, Greek society was in a process of rapid proletarianization, as in fact any other part of the capitalist world. This view was reflected on the party line and led to advocating a two-stage struggle: first achieving a people's republic and subsequently its transformation into a socialist republic.[4] A measure of working-class density (PROLET) is calculated

[4]The decisions was adopted by the 7th Party Congress, in October 1945; see KKE documents (1964, p. 534).

as the proportion of workers to total active population in each prefecture. A simple regression finds that class structure alone is insignificant, so that an instrumental variables approach, adopted as suggested by Miguel et al. (2004), appears to be more promising for conflict situations.

Proletarianization is expected to increase as a result of economic slackness, thus it must be positively correlated with the rate of unemployment. According to Marxist theory, proletarianization also increases by the accumulation of capital and this, in turn, increases the degree of production concentration.[5] An approximate measure of the concentration of production is the number of employees per employer; the latter being described by variable FIRMS. Both series of unemployment and production concentration are subsequently used as instrumental variables. Econometric results are displayed in Table 3.2 and show that working-class composition is now significant at the 10 % level. Nevertheless, the explanatory power with regards to the political discontent prevailing in 1946 is extremely low and the approach is not pursued any further.

Table 3.2 Class structure and political discontent[a] (standard errors in brackets)

		Instrumental variables[b]	
Independent variables ↓	Discontent	PROLET	Discontent
Method (S.E. in brackets)	OLS	OLS	2-Stage LS
Constant	21.278 (6.41) ***	−0.379 (4.14)	16.077 (7.73)**
Proletarianization[c] (workers as % of active population)	0.298 (0.22)	–	0.495 (0.28) *
Unemployment rate	–	3.847 (0.46) ***	–
Employees per firm	–	0.689 (0.27) **	–
Nobs	38	38	38
R^2 adj	0.019	0.651	−0.001
DW	0.93	1.33	0.90
F-stat (prob)	1.74 (0.195)	35.55 (0.00)	3.15 (0.084)

Notes: One, two or three stars indicate p-values at 10 %, 5 % or 1 % respectively. Data in Appendix
[a]Political discontent is measured as the simple average of abstention rates in the 1946 elections (ABS46) and the republican vote in the 1946 Referendum (REP46)
[b]Instrumental variables include the rate of unemployment and the degree of business concentration approximated by the ratio of employees (WORKF-UNEMP) to employers (FIRMS)
[c]Proletarianization (PROLET) is measured as the ratio of total workforce (WORKF) to active population (NACT)

[5]In his analysis of proletarianization, Karl Marx (1954) first asserts that "the amount of capital which is bound to be in the hands of each capitalist must keep increasing", and then concludes that the "[a]ccumulation of capital is, therefore, increase of the proletariat"; *Capital* (Ch. XIV and XXV, p. 340 and 576 respectively).

3.7.4 Entrenched Grievances

An alternative explanation is sought by attributing political discontent to long-entrenched problems that had caused extensive social rivalries in the past and further festered into bitter grievances during the Occupation.

Three such issues are considered: the first is related to the ethnic refugees who fled to Greece in 1922. To ensure their social and national integration, the Government financed several housing projects and included them in the land distribution program. The latter caused extensive friction and hostility among indigenous countrymen who saw the holdings initially promised to them being substantially reduced to accommodate newcomers. Moreover, refugees played an important role in radicalizing the working class movement, and several of them took up leading positions in the organization and expansion of the resistance formations during Occupation. As a result, the proportion of refugees to local population (variable REFG) is expected to be an important determinant of the intensity of political discontent.

The second issue is associated with the destruction of small and medium-size enterprises in the aftermath of the economic crisis in 1932. Small proprietors and businesses suffered further from the serial collapses of market exchange during Occupation and several of them turned to the guerrillas protest and also for protection. A measure of this grievance is obtained by the number of real-estate foreclosures enforced by the banks in 1934–1938, and expressed by variable FORCL.

A third factor of long-time discontent was stemming from the problems of rural populations. Small farmers had been particularly hit by the interwar crisis, as international demand was curtailed by the rising protectionism, while prices were collapsing by the steep devaluation of the currency.[6] During Occupation, farmers suffered from extensive confiscations of crops and livestock, and also from physical annihilation due to massive reprisals by the Occupation forces. A measure of rural discontent is provided by the index variable RUREV, which is a weighted average of pre-war rural revolts per prefecture according to their severity as has been constructed by Seferiadis (1999).

The above measures of long-held grievances are used as explanatory variables for political discontent, and estimates are shown in Table 3.3. All indices are now found to be statistically significant and correctly signed.

Grievance analysis suggests that political discontent in Greece during the 1940s was not a reflection of contemporaneous social conditions in general. Instead, social grievances were found to be strongly correlated with the share of agriculture in local economies, implying that the lack of development and regional integration gave rise to a structural reason for the poor living conditions. But even in the countryside, there was some hope that the situation would eventually improve. An

[6]Further details can be found in Christodoulakis (2013).

Table 3.3 Old grievances and political discontent

	Old grievances	
Independent variables ↓	Urban	Including rural
Method (S.E. in brackets)	OLS	OLS
Constant	14.462 (3.62)***	5.767 (5.44)
Refugees as % population	0.536 (0.19)***	0.852 (0.288)***
Prewar foreclosures per 100,000 inhabitants	0.538 (0.15)***	0.6205 (0.18)***
Index of agrarian revolts per 100,000 inhabitants	–	1.300 (0.66)*
Dummy for Messinia[a]	–	−30.88 (13.56)**
Nobs	36	26
R^2 adj	0.367	0.45
DW	1.26	1.88
F-stat (prob)	11.15 (0.0002)	6.11 (0.002)

Notes: Standard errors in brackets. One, two or three stars indicate p-values at 10 %, 5 % or 1 % respectively. Data as in Appendix

[a]A dummy is used for the prefecture of Messinia where the index of agrarian protest is unusually high, but political discontent too low—probably due to an overly rigged ballot

example in case is the problem of malaria that was ravaging rural populations until aid from the UN Relief Agency (UNRA) finally managed to contain it.

But in the absence of a political process that could guarantee reconciliation and participation in the prospective development, another critical issue was overlooked: As seen in Chap. 2, exclusion politics were imposed in Greece at a massive scale during the rightwing dictatorship 1936–1940, only to be brutally exceeded by the subsequent Axis Occupation. As suggested by Close (2003), old-entrenched grievances continued to fuel postwar rivalries. This soon led to a vicious circle: as the conflict was spreading, social conditions and exclusion tactics were further aggravated, leading to even more strife.

References

Alesina A, Perotti R (1996) Income distribution political instability, and investment. Eur Econ Rev 40(6):1203–1228. doi:10.1016/0014-2921(95)00030-5

Christodoulakis N (2013) Currency crisis and collapse in interwar Greece: predicament or policy failure? Eur Rev Econ Hist 17(3):272–293

Close DH (2003) The origins of the Greek Civil War. Filistor Editions (in Greek), Athens, Also Longman, London (1995)

Collier P, Hoeffler A (2004) Greed and grievance in Civil War. Oxf Econ Pap 56:563–595. doi:10.1093/oep/gpf064

Collier P, Sambanis N (2005) Understanding civil war: evidence and analysis, Volume 1. Africa. World Bank, Washington, DC, https://openknowledge.worldbank.org/handle/10986/7437

Datta S, Reimer J (2013) Malaria and economic development. Rev Dev Econ 17(1):1–15. doi:10.1111/rode.12011

Grossmann M, Lang M, Dietl H (2011) Transitional dynamics in a Tullock contest with a general cost function. BE J Theor Econ 11(1):1–26

Hirshleifer J (1989) Conflict and rent-seeking success functions: ratio vs. difference models of relative success. Public Choice 63(2):101–112

KKE (1964) Forty years 1918–1958: collection of documents. Party Edition, Athens (in Greek)

Konrad K, Skarpedas S (1998) Extortion economica. London Sch Econ Polit Sci 65(260):461–477

Mankiw NG, Phelps ES, Romer PM (1995) The growth of nations. Brook Pap Econ Act 1995(1): 275–326, URL: http://www.jstor.org/stable/2534576

Marx K (1954) Capital: a critique of political economy, vol I. Lawrence and Wishart, London

Miguel E, Satyanath S, Sergenti E (2004) Economic shocks and civil conflict: an instrumental variables approach. J Polit Econ 112(4):725–753

Seferiadis S (1999) Small rural ownership, subsistence agriculture, and peasant protest in interwar Greece: the agrarian question recast. J Mod Greek Stud 17(2), 277–323, October

Spence D, Hotchkiss J, Williams CS, Davies PD (1993) Tuberculosis and poverty. Br Med J 307(6907):759–761

Tullock G (1967) The welfare costs of tariffs, monopolies, and theft. West Econ J 5(3):224

Chapter 4
Models of Repression and Inclusion

The Collier-Hoeffler model is extended in two ways: first by examining the possibility of altogether avoiding the conflict if mutual concessions are agreed upon, and, second, to include the application of repression and reprisals by the government as deterrence for the guerrillas staging the rebellion. By allowing the fighting effectiveness to differ for the two sides, the effect of other factors such as terrain on the development of guerrilla forces is examined. Combinations of high repression and low inclusion in power sharing seem to be conducive to the onset of a civil war. Moreover, the application in the Greek civil war shows that persecutions tended to inflame the severity of the conflict, rather than suppress it.

4.1 Introduction

A number of extensions to the Collier-Hoeffler model are adopted so as to make it more realistic in capturing conflict situations. These include:

(a) The government is assumed to make a pre-emptive offer to the guerrillas in an attempt to forestall the conflict. It enters the conflict only if the offer is not accepted.
(b) The effect of repression and reprisals by the government is examined as part of its strategy to deter guerrillas from engaging in a civil war. Guerrillas are assumed to be unable to organize reprisals, since they lack state institutions and enforcement power. This feature creates asymmetries in the behaviour of the two sides and is used to explain several characteristics of civil wars that cannot be considered in the Collier-Hoeffler framework.
(c) The army threshold for the guerrillas entering the conflict is derived endogenously. Moreover, the fighting characteristics are allowed to differ across the two sides.

N. Christodoulakis, *An Economic Analysis of Conflicts*,
DOI 10.1007/978-3-319-32261-2_4

4.2 Repression and Inclusion

The notation and terminology used below is exactly the same to that used in Chap. 3, unless stated otherwise. The following additions are made:

4.2.1 Repression

In order to deter the conflict from the onset, the Government may adopt repression policies against prospective guerrillas in an attempt to diminish their expected payoffs and make their preparations more costly. Suppose that repression imposes a fixed cost of size (D) on the guerrillas. This may include job dismissals, prohibition of trade-union activity, prosecution of sympathizers, confiscation of land or other property, exile, and other forms of repression. By applying these measures, the government is assumed to derive a benefit (εD), where $0 < \varepsilon < 1$ denotes the degree of extraction through the imposition of repressive measures. Extraction gains are derived from various forms of rent-seeking created by the repressive regime or even from the direct appropriation of confiscated assets. The cost functions become:

$$C_1 = D + w_1 r \tag{4.1a}$$

$$C_2 = -\varepsilon D + w_2 q \tag{4.1b}$$

4.2.2 Reprisals

Guerrillas may be defeated with probability $(1-p_1)$, in which case they are likely to face reprisals in the form of being discriminated against or deprived of their future opportunities to earn an income. This is modelled by assuming that they receive only an outcome $L_1 < V_1$. Symmetrically if the state army is defeated with probability $(1-p_2)$, it gets $L_2 < V_2$.

4.2.3 Fighting Capabilities

Fighting effectiveness is assumed to differentiate across the two sides, reflecting different capabilities and training. Namely:

$$f_1(r) = A_1 r^{m_1} \text{ and } f_2(q) = A_2 q^{m_2} \tag{4.2}$$

For the guerrillas' army and state army respectively. The conflict success functions become:

$$p_1 = \frac{A_1 r^{m_1}}{A_1 r^{m_1} + A_2 q^{m_2}},\ p_2 = \frac{A_2 q^{m_2}}{A_1 r^{m_1} + A_2 q^{m_2}} \tag{4.3}$$

Parameters $(A_j, j = 1{,}2)$ denote conflict technology, while exponents (m_j) denote intensity in manpower and are expected to be higher the better trained is the respective army. In a civil war the state army is usually more professionally trained, better equipped and, additionally, might enjoy certain strategic advantages against the rebels. These assumptions are expressed by letting $m_1 < m_2$ and $A_1 < A_2$ respectively. Other cases can be similarly considered. The following relationship is established:

Proposition 1 Relative discriminatory power is equal to the ratio of the semi-elasticities of conflict success functions with regards to army size, i.e.

$$\left[\frac{\partial p_1}{\partial r/r}\right] \Big/ \left[\frac{\partial p_2}{\partial q/q}\right] = \frac{m_1}{m_2} \tag{4.4}$$

Proof is given in the Appendix. For $m_1 < m_2$, this property means that the guerrilla army's winning probability improves after an increase in its size, but to a lesser extent than an equi-proportional rise would have on the state army. The implication is that if the government has the resources to expand its army and still maintain a superior fighting capability, the winning probability of the guerrilla army would eventually diminish (in relative terms) even if a similar mobilization policy is adopted.

4.2.4 *Inclusion Offers*

In an attempt to thwart the outbreak of hostilities, the government is making an offer to guerrillas by proposing to cancel reprisals and persecutions, and include the protesting party in power sharing. If the guerrillas choose not to fight, they will receive a part (σV_2) of the government's endowment, with parameter $(\sigma < 1)$ denoting the degree of inclusiveness.

Under the above assumptions, the expected payoffs for the two opponents now depend on whether the civil war is avoided or indeed takes place. Two separate cases are described as follows:

Case I: Power-Sharing Without Civil War

$$\Pi_{10} = \sigma V_2 \tag{4.5a}$$
$$\Pi_{20} = (1-\sigma)V_2 \tag{4.5b}$$

Case II: Civil War, Repression and Reprisals

$$\Pi_1 = p_1V_1 + (1-p_1)L_1 - D - w_1r \tag{4.6a}$$
$$\Pi_2 = p_2V_2 + (1-p_2)L_2 + \varepsilon D - w_2q \tag{4.6b}$$

4.3 Avoiding the Conflict

4.3.1 Peace Chances

A civil war is avoided if and only if the conflict payoffs are inferior to the no-conflict outcomes for both parties, i.e. $\Pi_1 \leq \Pi_{10}$ and $\Pi_2 \leq \Pi_{20}$. They are written as:

$$p_1V_1 + (1-p_1)L_1 - D - w_1r \leq \sigma V_2 \tag{4.7a}$$
$$p_2V_2 + (1-p_2)L_2 + \varepsilon D - w_2q \leq (1-\sigma)V_2 \tag{4.7b}$$

Rearranging, two sufficient conditions are obtained as:

$$D + \sigma V_2 \geq H_1(r) \tag{4.8a}$$
$$\varepsilon D + \sigma V_2 \leq H_2(q) \tag{4.8b}$$

where functions (H_j, $j = 1,2$) are used for brevity and defined as:

$$H_1(r) := L_1 - w_1r + p_1 \cdot (V_1 - L_1) \tag{4.9a}$$
$$H_2\ (q) := w_2q + (1-p_2) \cdot (V_2 - L_2) \tag{4.9b}$$

Function $H_1(r)$ is the payoff of the guerrillas winning the conflict, net of the cost of repression. Function $H_2(q)$ is the payoff of the government losing the conflict. The two functions are used below to define under which circumstances the conflict may or may not be avoided. The following properties are established:

Proposition 2 Function $H_1(r)$ is concave and $H_2(q)$ is convex. Therefore, function $H_1(r)$ reaches a maximum value of H_{1max}, while $H_2(q)$ has a minimum value of H_{2min}.

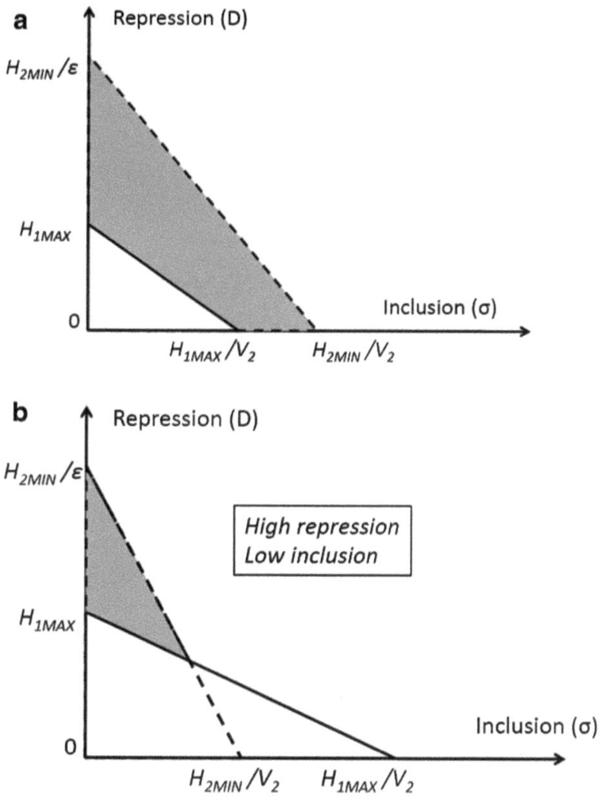

Fig. 4.1 (**a**) Likely peace. *Note:* Combinations of inclusion and repression to avoid a conflict are in the *shaded area*. (**b**) Unlikely peace. *Note:* The *shaded area* offers only a narrow combination of inclusion and repression

Proof is presented in the Appendix. To avoid the conflict, it suffices that the combination of repression and inclusiveness satisfies the following conditions:

$$D + \sigma V_2 > H_{1max} \tag{4.10a}$$

$$\varepsilon D + \sigma V_2 < H_{2min} \tag{4.10b}$$

Depending upon parameter values, the above constraints may or may not define a non-empty area in the (D, σ) policy space. An illustration of three possible cases is given in Fig. 4.1, and their policy implications are discussed below.

(a) *Feasible peace:* In the first panel, conditions (4.10a, 4.10b) define an area, within which any combination of repression and inclusion suffices to deter the civil conflict. By simple inspection, this is feasible if and only if

$$H_{1max} < H_{2min} \tag{4.11}$$

This condition is met when the guerrillas keep their expectations lower than what the government would receive in case is defeated. Condition (4.11) also ensures that repression is not too extractive (i.e. $\varepsilon < H_{2min}/H_{1max}$ or

$H_{1max} < H_{2min} / \varepsilon$). The feasible area offers a trade-off between repression and concession, where the conflict can be avoided. If the government is prepared to increase the degree of inclusion (i.e. a higher σ), repression can be reduced to lower levels, thus further re-enforcing the prospects of reconciliation between the two sides.

(b) *Unlikely peace:* In the second panel of Fig. 4.1, inclusion is low and an area of feasible combinations is found only by applying high repression. By simple inspection, this happens when

$$H_{2max} < H_{1max} < \frac{H_{2max}}{\varepsilon} \tag{4.12a}$$

In this case, the guerrillas have a higher band of expectations of what they might receive in case they win. The trade-off between the two policy options is much more restrictive, and only a choice of higher repression is likely to deter the conflict. This is in line with the findings by Hegre et al. (2001), who show that the incidence of civil wars increases with the degree of repression, unless it becomes too harsh. This option, however, may not be viable in the long run, as grievances will multiply and the decision to currently refrain from conflict may be overturned later on.

(c) *Impossible peace:* Finally, if the constraint lines are not intersected, no area of feasible combinations of repression and inclusion can be found. This happens when

$$H_{2max} < H_{1max} \text{ and } \varepsilon > H_{2max}/H_{1max} \tag{4.12b}$$

In this case, guerrillas are too demanding and repression by the government is too extractive so that a conflict is likely to erupt sooner or later.

4.3.2 Balance of Power

The cases of a peaceful settlement were discussed above in a static one-off setting. The implicit assumption is that the two adversaries consider entering a conflict against each other at a specific moment of history and, if they agree to settle their differences peacefully, the risk of confrontation is definitely over. This, however, may not be a realistic feature if adversaries have the opportunity to reconsider their case in the future. As noted by Konstantin and Schwarz (2008), a self-enforcing peace agreement between parties may not be viable if the transfer of resources and opportunities for alleviating the current conflict shifts the balance of power later on.

On the same topic, Gurr (1974) has noted that if a regime change takes place towards some more democratization, further changes may follow as the various actors adjust their possibilities and targets. In the present context, a shift may occur if the peace payoff in (4.5a, 4.5b) subsequently changes the relative fighting effectiveness in (4.2) in such a way that conditions (4.10a, 4.10b) no longer hold. Such an eventuality arises in the following cases:

1. The guerrillas initially accept the offer (4.5a) and then use the resources to increase their fighting capabilities. For example, access to, and participation in, high levels of power may enable the guerrillas to devise a better-informed strategy and/or attract new volunteers to their cause.
2. By transferring part of the assets (V_2), the government loses control of strategic resources and its capabilities are weakened. For example, by reallocating public revenues from military spending to social welfare in order to alleviate extensive grievances, the enforcement of security and safety is likely to be underfinanced.

In the face of a potential shift in the balance of power, the initial concession is not sustainable. Two possible solutions should be examined in this case: One is the signing of a long-term agreement ensuring that peace commitments are irrevocable. Another is to grant concessions in a repetitive way, so as conditions (4.10a, 4.10b) are subsequently satisfied as well.

4.4 Break-Out of Civil War

In the absence of options to adequately raise inclusiveness or apply repression, the eruption of civil war becomes more likely and the process is analyzed below by setting $\sigma = 0$. The government maintains repression *(D)* and expected payoffs are given as in (4.6a). The two sides need to organize armies so as to ensure that their conflict payoffs remain positive, i.e. $\Pi_j > 0,\ j = 1,2$. First the formation of the guerrillas' army is examined.

4.4.1 The Guerrilla Army

Substituting the conflict success function from (4.3) into (4.6a) and rearranging terms, the number of fighters *(r)* should be such that the following condition holds to ensure viability constraint $\Pi_1 > 0$:

$$w_1 r < -(D - L_1) + \frac{V_1 - L_1}{1 + (A_2/A_1) q^{m_2} r^{-m_1}} \tag{4.13}$$

The locus defined by the RHS expression is shown in Fig. 4.2 together with the LHS line.

The RHS is clearly increasing with the size of the guerrilla army *(r)*, and is easy to check that

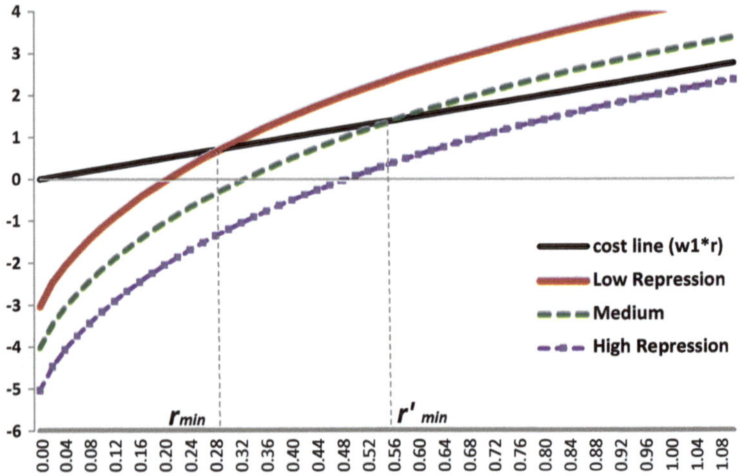

Fig. 4.2 The effect of repression on guerrillas' entry threshold. *Notes:* For illustration purposes, parameter values are set at $m_1 = 0.80$, $m_2 = 0.90$, $V_1 = 20$, $L_1 = 5$, $w_1 = 2.5$, $A_1 = 0.90$, $A_2 = 1$. Government army is normalized to $q = 1$. The low, medium and high reprisal levels are set at $D_1 = 8$, 9 and 10 respectively. For low reprisals, the threshold for entering the conflict is at $r_{MIN} = 0.28$, implying that a conflict may be viable only if guerrillas gather more than 28 % of the state forces. Raising repression to a medium level the cut-off rises to 0.58, while for higher levels there is no feasible threshold, and the conflict is abandoned

$$\lim_{r \to 0} RHS = -D + L_1 < 0 \tag{4.14a}$$

$$\lim_{r \to \infty} RHS = V_1 - D > 0 \tag{4.14b}$$

With low mobilization, the guerrillas risk losing the conflict outright and, in the face of extensive repression, they will choose not to fight, even in the absence of concessions from the government.

Proposition 3 For entering a conflict to be viable, the size of the guerrilla army should exceed a threshold that leads to positive payoffs:

$$r > r_{min} \tag{4.15}$$

The threshold is analytically derived in terms of model parameters, as described in the Appendix. The RHS expression in (4.13) is moving downwards as the level of reprisals *(D)* gets higher, making the entry condition more difficult to hold. Lower reprisals make rebellion easier to begin, as the entry threshold of mobilization is substantially reduced. If reprisals get tougher no cut-off threshold can be determined and rebellion is totally suppressed. A similar effect is observed as the state army increases in size *(q)* or receives better training (i.e. higher m_2).

It is straightforward to show that the RHS in (4.13) is moving upwards if conflict technology *(A_1)* is improving, and/or the victory reward *(V_1)* is increasing. Both lead to lower cut-off thresholds and thus make the entry condition easier to be satisfied. A similar effect occurs if manpower intensity (m_1) rises, due to better combat training and advanced mechanization of guerrilla fighting.

4.4.2 The State Army

Substituting the conflict success function from (4.16) into (4.17b) and rearranging terms, the size of the state army (q) should be such that the following viability condition holds:

$$w_2 q < (\varepsilon D + L_2) + \frac{V_2 - L_2}{1 + (A_1/A_2) r^1 q^{-m_2}} \tag{4.16}$$

The locus defined by the RHS expression is clearly increasing with the size of the state army (q), and is easy to check that

$$\lim_{q \to 0} RHS = \varepsilon D + L_2 > 0 \tag{4.17a}$$

$$\lim_{q \to \infty} RHS = V_2 + \varepsilon D \tag{4.17b}$$

The first limit implies that there is no threshold restriction for the size of the state army to fight in the civil war, no matter how low its mobilization might be. More details are given in the Appendix.

4.4.3 Optimal Armies

Solving the maximization problem for the system of two Eqs. (4.6a, 4.6b) the optimal response functions are obtained as follows:

$$r = \frac{m_1}{w_1} \cdot [V_1 - L_1] \cdot \frac{\partial p_1}{\partial r} \tag{4.18a}$$

$$q = \frac{m_2}{w_2} \cdot [V_2 - L_2] \cdot \frac{\partial p_2}{\partial q} \tag{4.18b}$$

For $m_1 < 1$ and $m_2 < 1$ it is easy to check that second-order derivatives are negative, thus the above response functions are concave and lead to the maximization of battle payoffs.

An interesting relationship between the sizes of the two armies is obtained. Taking partial derivatives of (4.3) as in the Appendix (Eqs. 4.21a, 4.21b) and dividing by parts we obtain that in equilibrium

$$\frac{r}{q} = \frac{m_1}{m_2} \cdot \frac{w_2}{w_1} \cdot \frac{V_1 - L_1}{V_2 - L_2} \tag{4.18c}$$

By using (4.3), the "win ratio" of the guerrillas is calculated as:

$$\frac{p_1}{p_2} = \frac{A_1}{A_2} \cdot \left[\frac{m_1}{m_2} \cdot \frac{w_2}{w_1} \cdot \frac{V_1 - L_1}{V_2 - L_2}\right]^{m_1} \cdot \frac{1}{q^{m_2 - m_1}} \tag{4.19}$$

These expressions are generalizations of the results obtained in the previous Chapter by using the Collier-Hoeffler model. They now include asymmetric fighting characteristics and the possibility of reprisals, thus they are capable of explaining more stylized facts in a civil war, in addition to those discussed in Sect. 3.5.

4.5 Stylized Facts of Conflict

By looking at the determinants of the win ratio in (4.19), the following issues are worth noting.

4.5.1 Counter-Productive Deprivation

An interesting remark can be made about the role of post-war treatment, which is expressed by the inferior outcome L_1 in case the guerrillas are defeated. In Sect. 5 it was shown that a higher level of repression *(D)* against the rebels raises the cut-off threshold and makes the start of the civil war less likely. However, if the government steps up deprivation by further cutting L_1, its effect may be counterproductive *after the civil war has erupted.*

A reduction in L_1 raises the win ratio for guerrillas and this happens because with more deprivation the opportunity cost of defeat is reduced. Realizing these prospects, more sympathizers decide to join the guerrilla army. The argument is in tune with the famous maxim put forward by Marx and Engels in their Communist Manifesto, according to which "*[t]he proletarians have nothing to lose but their chains. They have a world to win.*"[1]

[1]Marx and Engels (1978).

4.5.2 Capabilities and Terrain

The first ratio term in the RHS of (4.19) is the relative strength of fighting technology. It rises if guerrillas manage to acquire better and more arms, and falls if they are outpaced by the modern technology employed by the state army.

The same applies for the ratio of discriminatory powers (m_1/m_2). If guerrillas get inferior training (i.e. $m_1<m_2$) this reduces their chances to win the conflict. In that case, a rise in the size of the state army also reduces the win ratio for the guerrillas, as the expression becomes decreasing in variable (q).

In (4.19) the fighting effectiveness of the guerrillas (m_1) enters also as an exponent in the square-bracket expression. It is thus clear that the higher it is, the larger the effect of all other factors becomes in augmenting the win ratio. In practice this leads guerrillas to take maximum advantage of the terrain, in the hope that the state army cannot follow suit. For example, mountainous or densely forested areas are likely to increase the conflict technology of guerrillas relative to the state army and thus close the initial gap in fighting effectiveness.

4.5.3 Termination and Recurrence of Hostilities

The previous analysis can also be used to examine under which conditions the win probabilities get very low and guerrillas may choose to surrender after a defeat in the battle. If casualties in the specific battle are so high that the size of the guerrilla army falls below the cut-off threshold in (4.15), further fighting is rendered unrewarding. For the conflict to continue, guerrillas have to adequately meet one or more of the following requirements:

(a) Organize massive new recruitments to offset battle losses, thus ensuring again that their force satisfies constraint (4.15).
(b) Acquire better training and/or superior fighting skills (i.e. higher m_1 and higher A_1 respectively). For example, finance or intervention from abroad may provide new military equipment.
(c) If the two aforementioned possibilities fail, the only remaining option is to promise a higher victory reward either by contract or by means of propaganda. This explains why in many cases insurgents faced with the prospect of defeat become more sectarian and adopt a more extreme ideology, in the hope that they galvanize their remaining fighters.

4.5.4 Agendas and Propaganda Wars

The third ratio within the square bracket in (4.19) represents the net reward that is expected to be achieved by the rebels relative to that of the government. This is exploited by the insurgents, in order to act as a propaganda incentive for joining the struggle, as exact information on the true values may be missing. When a conflict erupts, the endowment V_2 that is currently controlled by the government is known to both parties, but the values of V_1, L_1 and L_2 are not known with certainty. The two sides may use (or abuse) evidence from other similar cases as a demonstration of their promises, in order to convince their audiences and enhance their win ratio.

For example, guerrillas frequently claim that, if they win, the economy will be boosted to a level V_1, which is bound to be higher than the current state V_2. Symmetrically, the government scorns the guerrillas' ability to build a better state, arguing that V_1 will collapse under their management. In this framework, several agendas that are put forward by the insurgent party against the incumbent in power can be discussed. For example:

An insurgency that originates from social and political exclusion due to ethnicity or other types of segregation hopes to mobilize supporters by promising participation in the process of distributing the reward (i.e. a high V1). A revolutionary ideology not only promises that rewards are higher in case of the insurgents' victory, but also that opportunities will open up for all members of society and not just for party sympathizers, also increasing the appeal of insurgency. This is captured by the fact that a higher L2 increases the win ratio. Even an insurgency advocated on religious grounds can be conveniently portrayed in terms of (4.19). In this case, the ratio of the two net rewards may serve to indicate the superiority of the value system of the insurgents versus that of the status quo.

4.6 APPLICATION: Repression and Inclusion in Interwar Europe

The previous analysis aimed to conceptualize the way in which inclusion and repression may influence the onset of a civil war. Some historical precedents can be used to provide an indicative confirmation as well.

4.6.1 Civil Wars in Europe

In the first half of the twentieth century Europe witnessed several civil wars and revolts. There have been extensive civil wars in Finland (1918), Hungary (1919), Ireland (1922), Spain (1936), Greece (1945 and 1946), Poland (1942), and Italy (1943). Additionally, the Baltic wars that took place in 1918 had been partly fought against domestic forces; see Balkelis (2015). In the beginning of the century there was also a massive peasant revolt in Romania (1907) against landowners, in which thousands of villagers perished; for an account, see Chirot and Ragin (1975). Less well-known is the fact that Belgium came on the brink of a civil war in 1944, but open conflict was finally avoided; see Conway (2012).

All of those conflicts had clear economic, national, or ideological agendas, thus they fit more easily into the framework of political calculation, as examined previously. It would, therefore, be interesting to compare those countries with the rest of European nations in terms of inclusion and repression.

In the absence of data explicitly measuring such characteristics, two approximate indices are used. In the empirical literature, education enrollment is found to have a strong negative correlation with the incidence and severity of civil wars. For example, Collier and Hoeffler (2004) find that a high index of male secondary school attendance significantly reduces the risk of conflict. Moreover, enrollment is found to have a strong positive relationship with per capita income. This is explained by the fact that more school-years typically lead to professional advancement, and, therefore, more income opportunities and higher social inclusion. For these reasons, a measure of inclusion is approximated by the average years of education in various countries during the interwar period, as shown in Table 11.4 in the Data Appendix.

On the other hand, a crucial form of repression in interwar Europe was manifested by imposing restrictions on trade union activity or banning membership outright. Hence, an index of repression has to be inversely related to the extent of trade unionism in total population. Juxtaposing the two proxies in Fig. 4.3, an interesting categorization emerges between countries that did or did not experience a civil war: countries with a higher inclusion did not experience a civil war, while all conflict countries were at that period characterized by relatively low inclusion and higher repression rates. Tellingly, Belgium emerges as a borderline case.

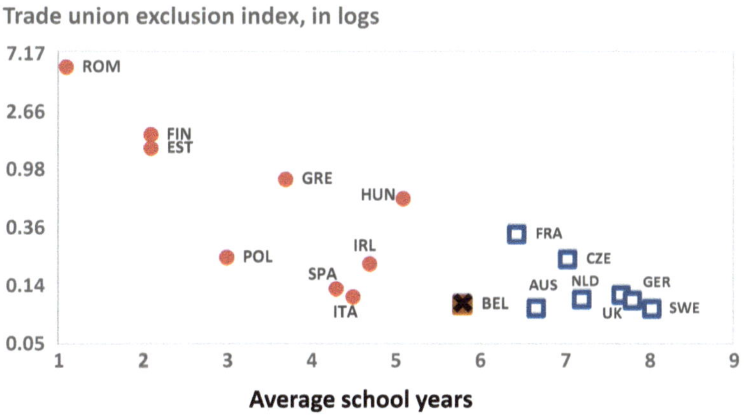

Fig. 4.3 Combinations of inclusion and repression. *Notes:* Average years of education in 17 European countries versus an index of non-participation in trade unions in the interwar period. *Source:* Data as in Table 11.4 (Appendix). The repression index is obtained as the simple inverse of union members/total population, in logarithmic scale

One should, of course, bear in mind that the above findings in no way suggest any kind of causation. For example, the high inclusion rates of Germany did not prevent it from falling into the even worse abyss of Nazism. The only message of the analysis is that a combination of high repression and low inclusion makes the onset of a civil conflict more likely.

4.6.2 Reconciliation and Reconstruction

Inclusion versus repression analysis can be used in a similar manner to analyze the options available to the winner after the conflict is terminated and the winning side faces the daunting task of rebuilding the state. As noted by van Boeschoten (2015), "two options are available for accomplishing this task: to proceed either by inclusion or by exclusion of the former internal enemies." Applying a policy of outright exclusion or low inclusion will necessitate a high degree of repression, and this leads to a continuation of grievances, thus fermenting the conditions of future conflicts.

European history of the twentieth century is not short of examples: In a recent study edited by Kissane (2015) on the reconciliation policies applied after civil wars in Europe, a whole range of combinations is found: The end of civil wars in Spain and Greece was followed by authoritative regimes and exclusion politics that sparked animosity and the recurrence of various types of political violence. In other countries, such as interwar Ireland, a more consensual process was applied in the post-conflict years, and this helped to reduce the legacy of hatred and social divisions in the newly established republic.

4.7 APPLICATION: Exclusion and Persecution in Greece

The combination of low inclusion and high repression that characterized Greece in the interwar period as shown in the left-upper quarter of Fig. 4.3, continued to rule even after the Occupation ended. In the absence of a genuine democratic process to express the new social balances that emerged after WW2, the previous political segregation ignited further tensions. The situation looked as dominated by "the winner takes all" syndrome, as discussed in Skaperdas (2008). Both sides seemed to be increasingly in favor of using force to exclude their opponents from participating in power-sharing or other distributional activities. Greece was quickly found to be in the area of low-inclusion and high-repression combinations, and an all-out confrontation was looming.

4.7.1 *Measuring Repression Politics*

A measure of political grievance is obtained by the number of persecutions *(PRSC)* initiated by the state against local militants. In the first phase of the conflict, such purges during 1945–1946 had been documented in a DAG Report submitted to the United Nations in 1948; see Grammenos and Rodakis (1947).

In the beginning of 1947, the Communist Party and other affiliated organizations were outlawed, and a massive purge of militants swept the country.[2] Emergency court procedures became swifter and stiffer; as a result prosecutions doubled in 1948 and death penalties increased threefold, as manifested in Fig. 4.4. A measure of political grievances in the period 1949:03–1949:08 is the number of citizens prosecuted during the previous years 1946–1948 in martial courts of nearby towns; for a documentation, see Michiotis (2007). Details of data definitions and regional classifications so as to correspond to the positions of DAG HQs are given in the Data Appendix (Table 11.3).

4.7.2 *Regional Formations*

After partisan warfare was adopted as the main form of the armed struggle, Greek guerrillas were gathering in the mountains and set up regional headquarters. Data on the guerrilla forces are available for 11 HQs in January 1948, and 21 HQs between March and August 1949. These data are used to analyze the spatial characteristics of conflict in terms of terrain and local political grievances in the two phases of the war. Note, however, that the small number of observations in 1948 makes the results for the first period to be only indicative.

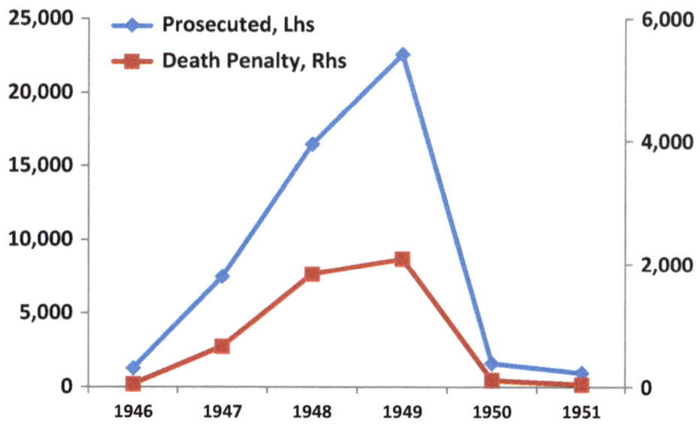

Fig. 4.4 Total prosecutions and death penalties. *Source:* Michiotis (2007, Tables 1 and 2, pp. 235–236)

[2]For a description, see Rizospastis (2011, p. 292).

Terrain is measured by the altitude (*MOUNT*) of mountains on which HQs were set up, and their distance (*DISTNB*) from the northern border of Greece. Other terrain indices, such as forest density, land cultivation, or country roads, were not found to be statistically significant and are not reported.

The following equation for regional DAG concentrations is finally estimated:

$$DAG^j = c^j + \beta_1 PRSC^j + \beta_2 MOUNT^j + \beta_3 DISTNB^j \tag{4.20}$$

Cross-section variables are per headquarter j ($j=1\ldots11$ in the first stage and $j=1\ldots21$ in the second). Constants (c^j) denote region-specific fixed effects. The sign of estimated coefficients depends on the form of the armed struggle and the relationship between the guerrillas and neighboring states. Thus, (β_1) would be negative if repression was effective in curtailing DAG recruitment and positive if it was counterproductive, as discussed in Sect. 5.2 as above. Coefficient (β_2) is expected to be positive for partisan warfare and negative if the struggle was mainly urban.

Finally, (β_3) must be negative if guerrillas were seeking refuge in neighboring states, and positive if the latter were allied with the Government and insurgents were handed over. The cross-section estimates are shown in Table 4.1 and some interesting findings are revealed.

A strong positive influence of prosecutions on recruitment is detected in both phases. This finding confirms Sambanis (2002), in claiming that political repression is usually counterproductive. The coefficients suggest that nearly half as many of those being persecuted chose to join DAG and thus avoid further purges. Regarding terrain, the results show that guerrilla concentrations were stronger in high mountains, in line with the partisan character of the conflict.

Table 4.1 Guerrilla formations and persecutions

Independent variable ↓	Guerrillas' spatial concentration	
	1948:01	1949:03–1949:07
Constant	−5839.79** (1862.03, p=0.014)	−1288.33* (672.6, p=0.06)
Local persecutions 1945–1946	0.497** (0.171, p=0.019)	
Court-martial prosecutions 7/1946–12/1948		0.429** (0.198, p=0.033)
Mountain altitude (m)	3.090** (1.066, p=0.015)	1.128*** (0.383, p=0.004)
Distance from borders (km)	–	−2.409*** (0.692, p=0.0007)
Method	OLS	OLS
Nobs	11 × 1	21 × 5
R^2 adj	0.799	0.275
F-stat	20.97 (p = 0.0006)	14.19 (p = 0.00)

Note: Standard errors in brackets. One, two or three stars indicate p-values at 10 %, 5 % or 1 % respectively. Data as defined in Appendix and Tables 11.2, 11.3

It is noticeable that the distance from northern borders is not found to be significant in 1948, as the strategy of DAG at that time was still to promote the conflict all over the country. The pattern changed in 1949, as DAG fighters retreated in the northern part of Greece and sought refuge in the communist Balkan states, thus the border effect is found to be negative and significant.

4.7.3 *Raising the Payoffs*

From the onset of the civil war, the Communist Party was lacking a clear strategy on the final objective and was opportunistically changing its meaning according to the developments in the battlefields. Initially, the aim was "democratic integration" so as to appeal to a wide audience. But as the prospects of winning the struggle started fading, the Party endorsed the "socialist transformation" of society as its ultimate objective. The move seemed irrational as it was undercutting any remaining political support from non-communist sympathizers, but is well explained by the discussion in Sect. 5.1. A Soviet-type transformation of the economy would involve large-scale confiscation of productive assets that would be placed in the hands of the Party. In this way, the reward (V_1) for those remaining in the Party ranks would be enhanced and so was hoped to be the case with the win ratio.

A major demonstration of such opportunistic tactics was the thesis adopted by the guerrilla leadership on the "national issue" regarding the treatment of the Greek-Macedonian populations in Northern Greece. In the aftermath of WW2, the Communist Party had initially adopted an uncompromising patriotic position, by pledging that the rights and equality of Slavic-speaking population should be implemented within the territory of Greece, for which "the present borders ... are sacrosanct and inviolable."[3] But in the beginning of 1949, civil war intensified and DAG Slav-Macedonian guerrillas were about to desert the ranks. To keep them close, the Party declared that they would be granted full "national emancipation"[4] or, in other words, that the prospective reward (V_1) would fall into their hands. But the move was counter-productive, as the Government immediately seized the opportunity and accused guerrillas for treason. It was only after its defeat that KKE admitted that it committed a "serious error."[5]

[3]Second Plenary, February 1946; reprinted in KKE (1964, p. 549).

[4]Fifth Plenary, January 1949; reprinted in KKE (1964, p. 577).

[5]Sixth Plenary, October 1949; reprinted in KKE (1964, p. 583).

Appendix

Proof of Proposition 1

Differentiating (4.3) we get

$$\frac{\partial p_1}{\partial r} = \frac{m_1}{r} \cdot \frac{A_1 A_2 r^{m_1} q^{m_2}}{\left[A_1 r^{m_1} + A_2 q^{m_2}\right]^2} = \frac{m_1}{r} \cdot p_1 \cdot p_2 \tag{4.21a}$$

Similarly

$$\frac{\partial p_2}{\partial q} = \frac{m_2}{q} \cdot \frac{A_1 A_2 r^{m_1} q^{m_2}}{\left[A_1 r^{m_1} + A_2 q^{m_2}\right]^2} = \frac{m_2}{q} \cdot p_1 \cdot p_2 \tag{4.21b}$$

Dividing (4.21a, 4.21b) by parts, expression (4.4) is readily obtained.

Proof of Proposition 2

Recalling (4.9a, 4.9b), taking into account (4.3), the concavity of $(p_j, j = 1,2)$ and the fact that $(L_j < V_j, j = 1,2)$ we obtain:

$$\frac{\partial^2 H_1}{\partial r^2} = \frac{\partial^2 p_1}{\partial r^2} \cdot (V_1 - L_1) < 0 \tag{4.22a}$$

and

$$\frac{\partial^2 H_2}{\partial q^2} = -\frac{\partial^2 p_2}{\partial q^2} \cdot (V_2 - L_2) > 0 \tag{4.22b}$$

Proof of Proposition 3

The first-order condition for maximizing (Π_1) as in (4.6a) yields that guerrillas should mobilize an army of a size such that it satisfies

$$w_1 = (V_1 - L_1) \cdot \frac{\partial p_1}{\partial r} \tag{4.23a}$$

Substituting from (4.21a) it gives that

$$w_1 r = m_1 \cdot (V_1 - L_1) \cdot p_1 \cdot (1 - p_1) \tag{4.23b}$$

The maximum payoff becomes

$$\Pi_{1max} = (V_1 - L_1) \cdot \left[m_1 p_1^2 + (1 - m_1) p_1 - \frac{D - L_1}{V_1 - L_1} \right] \tag{4.24}$$

For a repression strong enough so that $D > L_1$, the second-order expression within the square brackets has only one positive root, say $\pi_1 > 0$. In terms of model parameters it becomes:

$$\pi_1 = -\frac{1 - m_1}{2m_1} + \frac{1}{2m_1} \cdot \left\{ (1 - m_1)^2 + 4 \frac{D - L_1}{V_1 - L_1} \right\}^{1/2} \tag{4.25}$$

For the maximum payoff to be positive, and thus entering the conflict to be meaningful, it requires that $p_1 > \pi_1$. Given that probability p_1 is increasing in *(r)*, this condition is translated into an army threshold of $r > r_{MIN}$. The threshold in (4.15) is calculated in terms of model parameters from the inverse conflict success function:

$$r_{MIN} = inv\, p_1 [\pi_1] \tag{4.26}$$

Similarly the government's maximum payoff is obtained as

$$\Pi_{2max} = (V_2 - L_2) \cdot \left[m_2 p_2^2 + (1 - m_2) p_2 + \frac{L_2}{V_2 - L_2} \right] \tag{4.27}$$

This remains always positive; therefore, no threshold for the state army applies.

References

Balkelis T (2015) War, revolution and terror in the Baltic states and Finland after the Great War. J Baltic Stud 46(1):1–9. doi:10.1080/01629778.2015.1009685

Chirot D, Ragin C (1975) The market, tradition and peasant rebellion: the case of Romania in 1907. Am Sociol Rev 40(4):428–444

Collier P, Hoeffler A (2004) Greed and grievance in civil war. Oxf Econ Pap 56:563–595. doi:10.1093/oep/gpf064

Conway M (2012) The sorrows of Belgium: liberation and political reconstruction 1944–1947. Oxford University Press, Oxford

Grammenos B, Rodakis P (1947) That's how civil war started: reproduction of "The DAG Report to the UN". Glaros Editions, Athens (in Greek)

Gurr TR (1974) Persistence and change in political systems 1800–1971. Am Polit Sci Rev 68: 1482–15047

Hegre H, Ellingsen T, Gates S, Gleditsch NP (2001) Toward a democratic civil peace? democracy, political change, and civil war, 1816–1992. Am Polit Sci Rev 95:33–48

Kissane B (2015) After civil war. University of Pennsylvania Press, Philadelphia, PA

KKE (1964) Forty years 1918–1958: collection of documents. Communist Party edition, Athens (in Greek)
Konstantin S, Schwarz M (2008) A theory of brinkmanship, conflicts, and commitments. J Law Econ Org 24(1):163–183
Marx K, Engels F (1978) The communist manifesto. In: Marx K (ed) The revolutions of 1848. Penguin, London
Michiotis N (2007) In the name of the king: emergency martial courts in Greece 1946–1960. Synchroni Epochi Editions, Athens (in Greek)
Rizospastis (2011) The three-year epic of the Democratic Army of Greece 1946–1949. Synchroni Epochi Editions, Athens (in Greek)
Sambanis N (2002) A review of recent advances and future directions in the quantitative literature on civil war. Def Peace Econ 13(3):215–243
Skaperdas S (2008) An economic approach to analyzing civil wars. Econ Gov 9:25–44. doi:10.1007/s10101-007-0043-2
van Boeschoten R (2015) Enemies of the Nation-A nation of enemies: the long Greek civil war. In: Kissane B (ed) After civil war. University of Pennsylvania Press, Philadelphia, PA

Chapter 5
Dynamic Models of Conflict

This chapter provides a review of traditional models that are widely used for studying the dynamics of fighting. They include the combat models derived by Lanchester and the Lotka-Volterra models.

Though convenient in capturing various conflict interactions, their structure is not always adequate to describe the reality of actual hostilities. To overcome such shortcomings, an extended Lotka-Volterra framework is setup as a reaction function in Nash-Cournot competitions. The results provide a link between the dynamic systems of conflict and the opportunity-cost models derived in the previous chapters.

5.1 Modelling Conflict Interactions

Both the Lanchester variants and the family of Lotka-Volterra models describe dynamic interactions between populations by using differential equations. The main difference between the two model pedigrees is their origin: the Lanchester framework was setup to examine the ferocious battles of the First World War, while the Lotka-Volterra models were derived a little later, in order to study aggressive biological populations. Variants of the original models have been adopted to depict conflict situations in general, such as, for example, arms race, class war, and civil war. A long list of socio-economic, political and medical applications of such models can be found in Epstein (1997) and some of the most well-known cases are discussed below.

N. Christodoulakis, *An Economic Analysis of Conflicts*,
DOI 10.1007/978-3-319-32261-2_5

5.2 The Lanchester Model

Perhaps the most popular model in combat literature is the dynamic system of battle interactions between two adversaries that was derived by Lanchester (1916). Changing the wording of the original model so as to conform to the present configuration of a civil war, the adversaries are assumed to be the guerrillas and state army, controlling war resources of size (r) and (q) respectively. These variables may express the size of the army force, military equipment or other forms of fighting instruments. Using a system of differential equations, the two resources are depleted according to the laws:

$$dr/dt = -h_1 \cdot q \tag{5.1a}$$

$$dq/dt = -h_2 \cdot r \tag{5.1b}$$

where (h_1, h_2) denote combat effectiveness of the enemy force. The initial model has been extended in several ways in order to portray more realistic combat situations. For example:

- Combat parameters may be time-varying to express changing patterns of effectiveness, i.e. $h_1 = h_1(t)$ and $h_2 = h_2(t)$.
- A transformation of variables (r, q) into $q = \widehat{q} - z_1, r = \widehat{r} - z_2$ allows the Eqs. (5.1a, 5.1b) to be re-written as:

$$d\widehat{r}/dt = (h_1 z_1) - h_1 \cdot \widehat{q} \tag{5.2a}$$

$$d\widehat{q}/dt = (h_2 z_2) - h_2 \cdot \widehat{r} \tag{5.2b}$$

  This introduces own attrition rates $\left(h_j z_j, j = 1, 2\right)$ into the system.
- A multiplicative term (rq) may be added in the RHS of one or both the differential equations to denote the case where losses also depend on the dispersion of forces. With (s_1, s_2) denoting engagement constants, the extended model is written

$$dr/dt = -h_1 \cdot q - s_1 \cdot (rq) \tag{5.3a}$$

$$dq/dt = -h_2 \cdot r - s_2 \cdot (rq) \tag{5.3b}$$

  The linear terms denote the so-called "ancient battle" effect that is due to individual engagement in fighting. The multiplicative terms represent the "area fire" effect, when attrition is caused by shooting into an area rather than targeting an individual. As noted by Artelli and Deckro (2008), the "area fire" feature characterizes the impact of both artillery and shooting in the direction of a non-tracked target, and is a common feature in civil wars.

5.2.1 Congestion Effects

An interesting extension of the Lanchester model is the functional form introduced by Helmbold (1965) to capture congestion effects that limit battle effectiveness. The reason is that a rise in the relative size of one army makes maneuvering in the battlefield more difficult and raises this army's own losses. The combat model is set up as:

$$dr/dt = -h_1 \cdot q \cdot G_1(r/q) \tag{5.4a}$$

$$dq/dt = -h_2 \cdot r \cdot G_2(q/r) \tag{5.4b}$$

Congestion functions $G_j(.)$ are positive and increasing function, i.e. $G_j{}' > 0$, $j = 1,2$. Following Taylor (1981), one can express congestion effects by simple and separable power functions, as follows:

$$G_1(x) = x^a \, \text{and} \, G_2(x) = x^b \tag{5.5}$$

with $a, b > 0$. Substituting into (5.4a, 5.4b) and dividing by parts it is trivial to show that

$$\frac{dr}{dq} = \frac{h_1}{h_2} \cdot \frac{q^{1-a-b}}{r^{1-a-b}} \tag{5.6}$$

Solving the differential equation, the locus that interrelates the two armies is obtained as:

$$\frac{r^{2-a-b} + c_1}{q^{2-a-b} + c_2} = \frac{h_1}{h_2} \tag{5.7}$$

where c_1, c_2 are integration constants determined by initial conditions. For $a = b = 0$, the solution gives the so called Lanchester's "square law" as:

$$\frac{r^2 + c_1}{q^2 + c_2} = \frac{h_1}{h_2} \tag{5.8}$$

The "linear law" is obtained by any combination such that $a + b = 1$ and gives:

$$\frac{r + c_1}{q + c_2} = \frac{h_1}{h_2} \tag{5.9}$$

Notice that the linear law has essentially the same form as the relationship of army sizes derived in the Collier-Hoeffler context in Eq. (3.14a).

Finally, for $a = b = 1$, the "logarithmic law" is obtained for the two armies as:

$$\ln[r] = \frac{h_1}{h_2}\ln[q] + \frac{1}{h_2}\cdot[c_2h_1 - c_1h_2] \tag{5.10}$$

The above relationship implies that a change in government resources is matched by a proportionally change in guerrilla resources, scaled by their relative combat effectiveness (h_1/h_2).

5.2.2 Criticism

Despite the popularity of the Lanchester models, there are a number of shortcomings that limit their applicability. Lepingwell (1987) examined the realism of the Lanchester models and concluded that empirical data do not usually fit well with them. Nevertheless, he admitted that they remain popular because of their logical simplicity and theoretical elegance, which make them useful in conceptualizing a combat.

Another criticism stems from the deterministic context of the model. This implies that in every specific period of time a certain loss occurs in both armies, no matter whether or how a battle is indeed fought. For example, several combat variables are not continuous—as assumed by the differential model—and this implies that they change aperiodically. To avoid this feature, the model has to be transformed into a stochastic framework, in which the probability density of a specific level of forces at each period of time is derived.

To this effect, Wallis (1968) derived the win probability through a set of differential equations and obtained solutions via Laplace transformation. Jaiswal (1987) introduced several more variables and derived probability densities for the levels of different weapons and army units, which are then used in computer simulations. Furthermore, Jaiswal et al. (1995) used the stochastic Lanchester model to probabilistically study the outcome of a battle, when the two adversaries adopt different termination rules regarding losses.

Another snag of Lanchester laws is common with the Collier-Hoeffler model. Both models are built on the assumption that the main variable of consideration in battle decision-making is the size of the army. Each side is, thus, assumed to choose the army size every time it plans to fight against its opponent. In practice, the size of the army is not much flexible in a short period of time, and adversaries prefer to calculate how much effort is optimal to devote to a particular battle.

5.3 Lotka-Volterra Models

Lotka-Volterra models have also been employed to study human rivalry with variables expressing surviving populations or the outcome of adversarial actions, i.e. casualties, battle-deaths, injuries in riots, etc. Such models cover many areas: from political competition (e.g. Francisco 1996) to riots (e.g. Burbeck et al. 1978), and from civil conflicts (e.g. Francisco 2009) to outright revolution (e.g. Tsebelis and Sprague 1989). They are also used to relate population dynamics with the expansion of conflict behaviour. For example, in a historical study on the population process in ancient China, Turchin and Korotayev (2006) examine the effects of domestic warfare by using a Lotka-Volterra framework. Population and warfare are considered to be mutually competitive variables and their interactions are modelled in a predator–prey setting: population growth is negatively affected by warfare, while hostilities expand as population rises, because the number of violence practitioners increases.

5.3.1 *The Arms-Race Model*

One of the first attempts to model conflict-like competition was the framework developed by Richardson (1960) to study the interactions in the arms race between two countries. Richardson's initial model was written in continuous time and concerned the rate of change in the amount invested in armaments. Let (R) to denote the amount spent on arms by one nation and (Q) the respective amount spent by its adversary. The model is described as:

$$dR/dt = [\psi - \alpha R_t + \beta Q_t] \qquad (5.11a)$$

$$dQ/dt = [\omega - \gamma Q_t + \delta R_t] \qquad (5.11b)$$

If parameters (ψ, ω) are positive, they denote "grievances" against other nations that lead to armaments even in the absence of a similar policy by the adversary. In case they are negative they denote a "goodwill factor" towards disarmament. If parameters (α, γ) are positive they denote a saturation effect that becomes stronger at higher levels of armament spending. If positive, they imply expansionary pressures from within, due, for example, to the existence of a "military-industrial complex" that is influential in arms-related decision-making.[1] Finally positive parameters (β, δ) denote an escalation effect caused by the armaments of the adversary.

[1]The term was first used by US President Eisenhower in his Farewell Address, 21 January 1961.

Notice that the differential equations system of the Richardson's model is similar in structure with that described for the Lanchester models in the previous section, though the interpretation of parameters is different.

5.3.2 *Non-linear Models*

As with the simple Lanchester model, the Richardson equations cannot capture non-linear effects between two interacting parties. To that effect, a more useful framework is the Lotka-Volterra model, in which the evolution of each party depends on its own size, as well as on the interactions with the other.

Most applications of Lotka-Volterra model assume that the system is autonomous, i.e. free of external distrurbances. For simplicity, interactions are assumed to take place in continuous time, but discrete-time versions will follow later. With an over-dot denoting time differential and omitting time subscripts for simplicity, the autonomous Lotka-Volterra model for two interacting populations denoted by (R, Q) is written as:

$$\dot{R} = R \cdot [-\alpha + \beta Q] \tag{5.12a}$$

$$\dot{Q} = Q \cdot [-\gamma + \delta R] \tag{5.12b}$$

Parameters (α, β, γ, δ) may be interpreted according to the context they refer to, as explained below. Initially, Lotka-Volterra models were applied in biological applications, where variables represent species populations. If parameters are such that $\alpha < 0$, $\gamma < 0$, $\beta < 0$ and $\delta > 0$, the above equations describe a predator–prey system. Prey population (R) increases by the net fertility rate ($-\alpha > 0$) and is annihilated at a rate that is proportional to the predating population (Q) and its ferocity (β). Predators increase by fertility ($-\gamma > 0$) and the consumption of preys at a rate (δ).

Other types of behavior are easily portrayed. Interactions imply behavioral mutualism (or symbiosis) if parameters are such that $\alpha < 0$, $\gamma < 0$, $\beta > 0$, $\delta > 0$. Competition models of bilateral attrition are obtained by letting $\beta < 0$, $\delta < 0$, while own-decline is described by $\alpha > 0$, $\gamma > 0$. A comparative analysis of their properties can be found in Zhang et al. (2007).

5.3.3 *Steady-States and Stability*

By setting $\dot{R} = 0$ and $\dot{Q} = 0$, four steady states (R^*, Q^*) are easily obtained: one is the trivial equilibrium at $\Sigma_1 = (0, 0)$, and the other three are $\Sigma_2 = (0, \alpha/\beta)$, $\Sigma_3 = (\gamma/\delta, 0)$ and $\Sigma_4 = (\gamma/\delta, \alpha/\beta)$. The dynamics of the system around these equilibria are determined by the characteristic roots of the Jacobian matrix:

$$J = \begin{bmatrix} \frac{\partial \dot{R}}{\partial R} & \frac{\partial \dot{R}}{\partial Q} \\ \frac{\partial \dot{Q}}{\partial R} & \frac{\partial \dot{Q}}{\partial Q} \end{bmatrix} \tag{5.13}$$

The following Jacobian matrices are formed:

$$J_1 = \begin{bmatrix} -\alpha & 0 \\ 0 & -\gamma \end{bmatrix}, J_2 = \begin{bmatrix} -\alpha & \frac{\beta\gamma}{\delta} \\ 0 & 0 \end{bmatrix}, J_3 = \begin{bmatrix} 0 & 0 \\ \frac{\alpha\delta}{\beta} & -\gamma \end{bmatrix}, J_4 = \begin{bmatrix} 0 & \frac{\beta\gamma}{\delta} \\ \frac{\alpha\delta}{\beta} & 0 \end{bmatrix} \tag{5.14}$$

The characteristic roots of each matrix are obtained as the corresponding pairs $(-\alpha, -\gamma), (0, -\alpha), (-\gamma, 0)$ and $(\pm\sqrt{\alpha\gamma})$ respectively. In the first case of trivial equilibrium, stability holds if and only if $\alpha > 0$ and $\gamma > 0$. The second and third cases are indeterminate, as one characteristic root is zero. The fourth case with two non-trivial equilibria is *always* unstable for $\alpha\gamma > 0$, while a limit cycle occurs only if $\alpha\gamma < 0$.[2] This brief description of parameters, equilibria and stability has important implications when a Lotka-Volterra system is used to model a conflict situation.

5.3.4 Lagged Discrete-Time Models

Most of the coinflict applications are in discrete time in order to facilitate econometric estimation. The interacting populations are assumed to represent battle losses of the two adversaries: namely those of guerrillas (R) and the government (Q). In several applications with discrete-time Lotka-Volterra models it is assumed that current battle outcomes are determined by the size of the enemy lagged one period. An autonomous discrete-time system with lagged interactions between adversaries is written as:

$$\Delta R_t = R_{t-1} \cdot [-\alpha + \beta Q_{t-1}] \tag{5.15a}$$

$$\Delta Q_t = Q_{t-1} \cdot [-\gamma + \delta R_{t-1}] \tag{5.15b}$$

In the above expressions, guerrillas' losses in each period rise by an amount that is proportional to the battle effort exercised by the state army in the previous period. The increase in losses rises further if guerrillas had a high toll (R_{t-1}) in the previous battle, as this may have incapacitated their fighting ability. A similar pattern is assumed for the state army.

---

[2]Note that the product of the two characteristic roots equals the determinant of the Jacobian that in this case is given by det (J4) $= -\alpha\gamma.i$

A shortcoming of the autonomous models is that they do not have any stable non-trivial equilibrium, thus they are incapable of modelling possible conflict traps in a repetitive conflict. To show this, let $\Delta R = 0$ and $\Delta Q = 0$ to obtain four steady-states (R^*, Q^*), exactly equal to those in the continuous-time case. The dynamics of the system around these equilibria are determined by the characteristic roots of the Jacobian matrix:

$$J = \begin{bmatrix} \frac{\partial R_t}{\partial R_{t-1}} & \frac{\partial R_t}{\partial Q_{t-1}} \\ \frac{\partial Q_t}{\partial R_{t-1}} & \frac{\partial Q_t}{\partial Q_{t-1}} \end{bmatrix} \tag{5.16}$$

The four Jacobian matrices corresponding to the four equilibria are now:

$$J_1 = \begin{bmatrix} 1-\alpha & 0 \\ 0 & 1-\gamma \end{bmatrix}, J_2 = \begin{bmatrix} 1-\alpha & \frac{\beta\gamma}{\delta} \\ 0 & 1 \end{bmatrix}, J_3 = \begin{bmatrix} 1 & 0 \\ \frac{\alpha\delta}{\beta} & 1-\gamma \end{bmatrix}, J_4 = \begin{bmatrix} 1 & \frac{\beta\gamma}{\delta} \\ \frac{\alpha\delta}{\beta} & 1 \end{bmatrix} \tag{5.17}$$

The characteristic roots are respectively obtained as the pairs

$$\left(1-\alpha, 1-\gamma\right), \left(1-\alpha, 1\right), \left(1, 1-\gamma\right), \text{and} \left(1 \pm \sqrt{\alpha\gamma}\right) \tag{5.18}$$

For the system to be asymptotically stable, the characteristic roots must have a modulus within the unit disc, and four cases are examined in turn: In the case of trivial equilibrium, stability holds if and only if $0 < \alpha < 2$ and $0 < \gamma < 2$. The second and third cases in (5.18) are indeterminate. The fourth case corresponds to a non-trivial equilibrium but is *always* unstable, as one characteristic root clearly lies outside the unit disc. If $\alpha\gamma < 0$, then roots are complex, both lie outside the unit circle, and an unstable limit cycle occurs.

It is noticeable that the occurrence of a non-trivial limit cycle requires only that survival parameters α or γ take opposite signs, or, in other words, one of the armies becomes self-destructive and past casualties lead to more losses in the future. This might happen only in circumstances that the army is being constantly depleted from critical support units or is panic-stricken after a major defeat. Neither assumption is compatible with a prolongation of conflict.[3] But even in such an unlikely fighting case, for the non-zero steady-state ($R^* = \gamma/\delta$, $Q^* = \alpha/\beta$) to be meaningful and take

[3]Losses can also be self-multiplied in suicidal insurgencies where a new wave of martyrs follows those previously perished but, again, this cannot last for very long.

positive values, the condition $\alpha\gamma < 0$ requires in turn that $\beta\delta < 0$. Again this would imply improbable opposite behaviours for the two fighting sides.

Another snag in model (5.15a, 5.15b) is the use of lagged terms. This may be convenient in econometric estimation but is clearly unrealistic in portraying an actual battle process. Literally interpreted, it implies that a current battle outcome is exclusively determined by the outcome of the previous battle. But in such a case, adversaries could easily calculate the outcome and avoid going to a battle again. Therefore the lagged model is unsuitable for capturing real-life hostilities.

An example in case is the book by Francisco (2009) that analyzes civil conflict in Europe and elsewhere. By using an autonomous Lotka-Volterra framework with a lag, the model examines the interactions between the fighting populations. Upon closer examination, however, the interaction coefficients are found to render meaningless equilibrium values or improbable dynamics. Two examples of misleading applications of Lotka-Volterra models are demonstrated in Appendix 5B at the end of this chapter.

In conclusion, autonomous discrete-time systems preclude the existence of stable non-trivial steady states, thus they are unsuitable to describe prolonged conflicts.

5.3.5 *Fractional Models*

In other applications, a more complicated non-linear structure is adopted in order to obtain stable non-trivial equilibria. The fractional model takes the form[4]:

$$R_t = \frac{R_{t-1}}{1 + \eta R_{t-1}} \cdot [-\alpha + \beta Q_{t-1}] \tag{5.19a}$$

$$Q_t = \frac{Q_{t-1}}{1 + \varepsilon Q_{t-1}} \cdot [-\gamma + \delta R_{t-1}] \tag{5.19b}$$

This model is similar to that adopted by Din (2013) for competing populations. By setting $R_t = R_{t\text{-}1}$ and $Q_t = Q_{t\text{-}1}$, it is easily shown that the system has four equilibria at

$$\Sigma_1 = (0, 0)$$

[4]Din (2013) describes a predator–prey model, in which interactions differ from the conflict framework. The corresponding parameters should become α, $\gamma<0$ so that $(-\alpha)$ and $(-\gamma)$ denote reproduction rates. Parameters in the non-linear terms should be $\beta<0$ to denote predation, while $\delta>0$ to express the effect of feeding the predators.

$$\begin{aligned}\Sigma_2 &= \left(-\frac{1+\alpha}{\eta}, 0\right) \\ \Sigma_3 &= \left(0,\ -\frac{1+\gamma}{\varepsilon}\right) \\ \Sigma_4 &= \left\{\frac{(1+\alpha)\varepsilon+(1+\gamma)\beta}{\beta\delta-\eta\varepsilon}, \frac{(1+\alpha)\delta+(1+\gamma)\eta}{\beta\delta-\eta\varepsilon}\right\}\end{aligned} \tag{5.20}$$

Evaluating a similar Jacobian as in (5.16), Din (2013) shows that the non-trivial equilibrium (Σ_4) is asymptotically stable only if several complicated conditions are imposed upon the parameters.

5.3.6 *Criticism*

All the above models unrealistically assume that a prolonged conflict has only autonomous dynamics and remains immune from the external environment. This may suit to biological experiments in vitro, but hardly is a sensible assumption with regards to actual wars. Combined with the autonomous form, the use of discrete time renders most of those models either too complex for, or incapable of, capturing important characteristics of real-life conflicts.

A second snag is that in both the autonomous discrete-time framework with lagged interactions and the continuous-time system with contemporaneous terms, a limit cycle is obtained only if parameters (α, γ) are of opposite sign. As noted by Zhang et al. (2007), this happens in a situation with predator–prey patterns. But to assume that guerrillas (or the state army) are prepared to act as preys and still enter a civil war is rather awkward.

A third shortcoming of the traditional Lotka-Volterra models is that although they are used to describe conflicting behaviours, they lack a formal definition of such a conflict framework in the first place. The dynamic equations are ad hoc adaptations of the models used in the species literature, and their interpretation along battle characteristics is made ex post. Thus their comparison with the opportunity-cost models is not possible.

In order to setup a system where conflict traps may occur, the traditional Lotka-Volterra framework needs to be modified in three important aspects:

First, instead of making the unrealistic assumption that an army's current casualties are determined by the opponent's past losses, the model should reflect contemporaneous interactions between the two combatants, as happens in actual battlefields.

Second, it should allow for exogenous factors to influence both the path of events and the size and type of the steady state.

The third modification regards the existence of non-trivial conflict equilibria. In practice, it is more likely that a conflict is prolonged exactly because the two sides adopt similar rather than diverging fighting patterns. This asks for both parameters (α, γ) being similarly signed and is more suitable for modelling an organized civil strife.

Fourth, the dynamic equations need to be derived as reaction functions in a well-defined conflict framework as per below. The varieties discussed above can then be obtained as partial or reduced-form cases as described in the Appendix.

5.4 The Lotka-Volterra Model as a Cournot Competition

One way to overcome the ad hoc specifications is to try to derive the Lotka-Volterra equations as reaction functions in game-theoretic models, whereas each competing party chooses a strategy to maximize a specific payoff. The link between games of conflict and Lotka-Volterra models has been recognized since long ago. For example, the Richardson's arms-race model has formally been derived as a differential game between two adversaries with payoff functions and resource constrains; for details see Simaan and Cruz (1975). More recently, Linn (2005) described how Lotka-Volterra equations essentially are Cournot models with two competing populations. Below it is shown how the Lotka-Volterra equations can be derived.

5.4.1 Payoffs in a Lotka-Volterra Framework

Using the framework described in Chaps. 3 and 4, a civil war can be described as a deadly contest, in which the prize is victory in the battlefield. A simple and definite measure of the value of a battle victory is obviously the number of casualties inflicted on the opponent.

Considering the enemy's losses as a measure of successful battle outcomes is similar to the payoff function adopted by Wei et al. (2008). Moreover, it is in tune with famous war doctrines. No less than the renowned Prussian general and military theorist Carl von Clausewitz set the objective for an army in war by stating that "*the defeat and destruction of [its opponent's] fighting force remains the best way to begin and will in any case be a significant feature of the campaign*"; Clausewitz, (1976, p. 596).

In a similar vein, modern guerrilla theorist and practitioner Che Guevara (1961, Ch. 1) left no doubt about the ultimate aim of guerrilla operations. In an attempt to alleviate any hesitations about the true nature of revolting against oligarchic regimes in Third World countries, he emphatically stated that "*a [civil war] is always a struggle in which each contender tries to annihilate the other.*"

Moreover, modelling conflict in terms of battle casualties is likely to facilitate econometric estimation as well. Collier and Hoeffler (2007) suggest that by using battle casualties rather than army populations, a more reliable assessment of the conflict is obtained.

Let's start with the strategy of guerrillas and then proceed similarly with the state army. Suppose that a battle takes place in each period t ($t = 0, 1, 2, \ldots$) and the objective for the guerrillas is to inflict the maximum number of casualties (Q_t) on the state army, so as to reduce fighting capabilities and force the government to surrender. The effort invested in the battle is an increasing function of the sacrifices that the guerrillas themselves are prepared to suffer and can be measured by the number of their own casualties (R_t).

In the battle taking place at time period (t), the guerrillas choose their effort to maximize a net outcome as follows:

$$\max_{R_t} \Pi_{1t} = Q_t^e - C_1(R_t) \tag{5.21a}$$

Superscript (e) indicates the expected level of enemy losses. Similarly, the state army chooses the effort in terms of its own casualties, in order to maximize the net outcome:

$$\max_{Q_t} \Pi_{2t} = R_t^e - C_2(Q_t) \tag{5.21b}$$

In the above expressions, (C_j, $j = 1, 2$) are cost functions for the guerrillas and the state army respectively. They are assumed to be increasing in the number of own casualties and also in the rate at which such casualties rise from one period to another. If the rate is high enough, it implies that losses are accelerated and the fighting forces will soon be decimated. An example of battle-cost functions is the following:

$$C_1(R_t) = d_1 R_t + g_1 \cdot \left[\frac{R_t}{R_{t-1}} - 1\right] \tag{5.22a}$$

$$C_2(Q_t) = d_2 Q_t + g_2 \cdot \left[\frac{Q_t}{Q_{t-1}} - 1\right] \tag{5.22b}$$

Parameters $\left(d_j,\ g_j,\ j = 1,\ 2\right)$ are weighting factors.

Next, the function of the opponents' casualties is derived. The guerrillas assume that the level of government casualties will be determined by the effort that they are prepared to expend. The function of expected losses is assumed to take the form:

$$Q_t^e = A_1 \cdot [R_t - \theta_1]^{k_1} \tag{5.23a}$$

Similarly, the losses expected to be inflicted upon guerrillas by the state army depend on the latter's losses as follows:

$$R_t^e = A_2 \cdot [Q_t - \theta_2]^{k_2} \tag{5.23b}$$

In expressions (5.23a, 5.23b), the RHS represents a "war destruction" function. Parameters (A_j) denote the level of "war technology" as described in Chap. 4, while (k_j) are assumed to be below unity, thus destruction is characterized by decreasing returns on own losses. The rationale behind this condition is that if—for example—the number of guerrilla losses becomes excessive, some critical capabilities may collapse and their destructive effect is diminished. Finally, the effect of war effort is likely to materialize after the latter surpasses a certain threshold (θ_j, $j = 1, 2$). The more adverse is the external environment, the higher the threshold.

5.4.2 Reaction Functions

The first-order conditions for maximizing battle outcomes are obtained by differentiating (5.23a, 5.23b). Taking (5.21a, 5.21b) and (5.22a, 5.22b) into account, we arrive at the expressions:

$$\frac{\partial \Pi_{1t}}{\partial R_t} = 0 \Rightarrow \frac{k_1}{R_t - \theta_1} \cdot \left[Q_t^e\right] = d_1 + \frac{g_1}{R_{t-1}} \tag{5.24a}$$

$$\frac{\partial \Pi_{2t}}{\partial Q_t} = 0 \Rightarrow \frac{k_2}{Q_t - \theta_2} \cdot \left[R_t^e\right] = d_2 + \frac{g_2}{Q_{t-1}} \tag{5.24b}$$

The above expressions denote Nash-Cournot reaction functions. As the battle proceeds, the actual size of losses may differ from expected ones by an error margin, i.e.

$$R_t = R_t^e + e_1 \tag{5.25a}$$

$$Q_t = Q_t^e + e_2 \tag{5.25b}$$

Constants (e_j, $j = 1, 2$) represent losses that occur exogenously due to unforeseen battle developments for the guerrillas and the state army respectively.

Solving the above equations, actual casualties are obtained as follows:

$$R_t = \frac{k_1 R_{t-1}}{g_1 + d_1 R_{t-1}} \cdot [-e_2 + Q_t] + \theta_1 \tag{5.26a}$$

$$Q_t = \frac{k_2 Q_{t-1}}{g_2 + d_2 Q_{t-1}} \cdot [-e_1 + R_t] + \theta_2 \tag{5.26b}$$

Regarding the second-order conditions, it is easy to show that

$$\frac{\partial^2 \Pi_{1t}}{\partial R_t^2} < 0 \, \text{and} \, \frac{\partial^2 \Pi_{2t}}{\partial Q_t^2} < 0 \quad (5.26c)$$

as long as $k_1 < 1$ and $k_2 < 1$ respectively.

5.5 The Conflict Trap

Several ad hoc Lotka-Volterra models can be derived as special cases by appropriately setting the parameter values. In particular, a model with stable non-trivial equilibrium can be obtained by using the Cournot-Nash reaction functions.

5.5.1 A Non-autonomous Model

A non-autonomous discrete-time system is obtained by assuming away the proportional terms in cost functions (5.22a, 5.22b), i.e.

$$d_1 = d_2 = 0 \quad (5.27a)$$

For convenience, a new set of parameters is also defined as:

$$\alpha = 1 + e_2 \frac{k_1}{g_i}, \beta = \frac{k_1}{g_1}, \gamma = 1 + e_1 \frac{k_2}{g_2}, \delta = \frac{k_2}{g_2} \quad (5.27b)$$

The model in (5.26a, 5.26b) finally takes the form:

$$\Delta R_t = R_{t-1}[-\alpha + \beta Q_t] + \theta_1 \quad (5.28a)$$

$$\Delta Q_t = Q_{t-1}[-\gamma + \delta R_t] + \theta_i \quad (5.28b)$$

The Greek-letter parameters may be interpreted to reflect some interesting fighting characteristics. The first terms in the RHS square bracket of the above equations denote how the rise in casualties of each army is influenced by its own past losses: if parameters (α, γ) are positive, they denote "survival rates" (i.e. $-\alpha$, $-\gamma < 0$) as each army gradually learns how to contain fatalities by better training and defence-building. If negative they denote "own-attrition" rates (i.e. $-\alpha$, $-\gamma > i$) due to fatigue or loss of critical units and this is similar to the modified Lanchester model in Eqs. (5.3a, 5.3b).

The implication is that rebel losses are increasing in the intensification of the battle, which is assumed to be proportional to the casualties suffered by the aggressor. Parameters (β, δ) in the RHS express "firing effectiveness", as in the terminology introduced by Epstein (1997, Ch. 2). The effect is multiplied by past own-casualties, as these make the guerrillas to be more vulnerable. Parameter values may differ for the two sides because of different firing capabilities.

Constants ($\theta_j, j = 1, 2$) represent exogenous adverse factors for the guerrilla and state armies respectively. For example, rough terrain for untrained troops, insufficient medical care or deterioration of logistics may steadily raise the toll, even without commencing the battle. If negative, they indicate steady improvements. If the logistics of the state army happen to be more frequently replenished and outperform those of the guerrillas, it is expected that $\theta_1 > \theta_2$. In general, they may be time-varying parameters to signify how the external environment changes during the conflict.

5.5.2 *Conflict Equilibria*

A conflict trap is defined as a situation where fighting is recurring without any side gaining an advantage decisive enough so as to terminally capitulate its opponent. In terms of model (5.28a, 5.28b), such a stalemate is represented by an asymptotically stable and non-trivial positive equilibrium (R^*, Q^*). Setting $\Delta R = \Delta Q = 0$, rearranging and solving the system, we obtain the system of two non-linear equations:

$$Q^* = \frac{\alpha}{\beta} - \frac{\theta_1}{\beta R^*} \tag{5.29a}$$

$$Q^* = \frac{\theta_2}{\gamma - \delta R^*} \tag{5.29b}$$

A graphical display of (5.29a, 5.29b) in Fig. 5.1 shows that there might be up to two stalemate equilibria. To determine the conditions for existence, the following Propositions are easily established. Proofs are given in Appendix 5A at the end of this chapter.

Proposition 1 Two positive equilibria exist if and only if the combination of survival rates is either too low or too high, i.e. if one of the following conditions is satisfied:

$$\sqrt{\alpha\gamma} < \left|\sqrt{\beta\theta_2} - \sqrt{\delta\theta_1}\right| \tag{5.30a}$$

or

$$\sqrt{\alpha\gamma} > \sqrt{\beta\theta_2} + \sqrt{\delta\theta_1} \tag{5.30b}$$

The feasible areas are shown in Fig. 5.2.

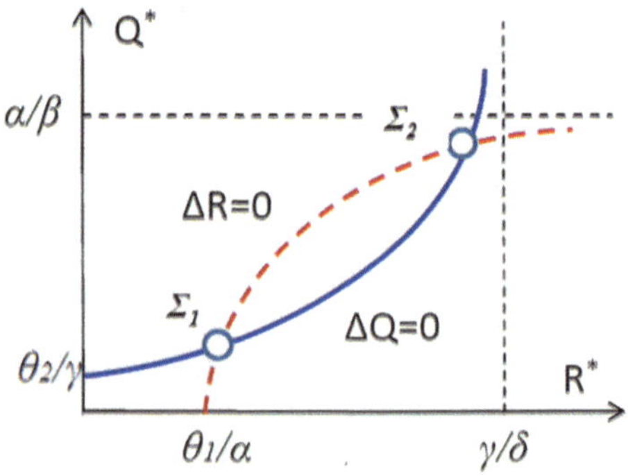

Fig. 5.1 Positive conflict equilibria

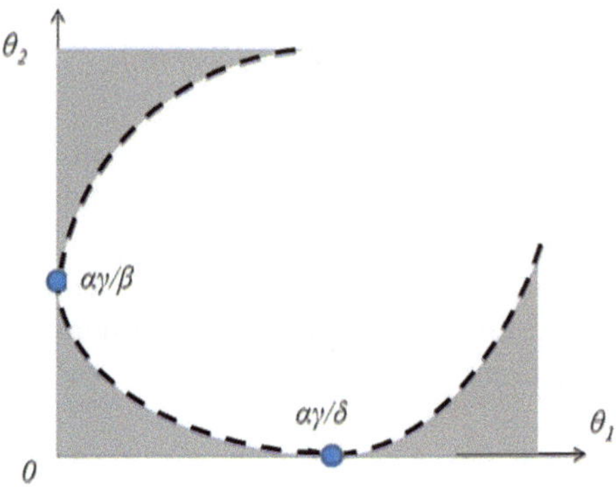

Fig. 5.2 Parameter areas for positive conflict equilibria

Proposition 2 No positive equilibrium exists if the survival rates lie within the intermediate range:

$$\left|\sqrt{\beta\theta_2} - \sqrt{\delta\theta_1}\right| < \sqrt{\alpha\gamma} < \sqrt{\beta\theta_2} + \sqrt{\delta\theta_1} \tag{5.31}$$

Proposition 3 A unique positive equilibrium exists if and only if one of the conditions (5.30a, 5.30b) holds as equality.

5.5.3 Stable Equilibria

The characteristics roots that drive the dynamics of the system are the eigenvalues of the Jacobian matrix evaluated at the steady-state (R^*, Q^*) as follows:

$$J(R^*, Q^*) = \begin{bmatrix} \frac{\partial R_t}{\partial R_{t-1}} & \frac{\partial R_t}{\partial Q_{t-1}} \\ \frac{\partial Q_t}{\partial R_{t-1}} & \frac{\partial Q_t}{\partial Q_{t-1}} \end{bmatrix} = \frac{1}{\Omega} \cdot \begin{bmatrix} 1 - \frac{\theta_1}{R^*} & \beta R^* \left(1 - \frac{\theta_2}{Q^*}\right) \\ \delta Q^* \left(1 - \frac{\theta_1}{R^*}\right) & 1 - \frac{\theta_2}{Q^*} \end{bmatrix} \tag{5.32}$$

with expression (Ω) defined as:

$$\Omega \triangleq 1 - \beta\delta R^* Q^* \tag{5.33}$$

Convergence to a stalemate equilibrium requires that the eigenvalues have a modulus below unity, while complex roots imply limit cycles.[5] However, expressions are non-linear and results can only be obtained numerically.

Appendix

A. Proof of Propositions 1, 2, 3

Combining (5.29a, 5.29b), the steady-state R^* is obtained as a positive root of the second-order equation:

$$\alpha\delta x^2 - (\alpha\gamma + \theta_1\delta - \theta_2\beta)x + \theta_1\gamma = 0 \tag{5.34}$$

Let us denote the discriminant as:

$$\mathfrak{D} = (\alpha\gamma + \theta_1\delta - \theta_2\beta)^2 - 4\alpha\gamma\delta\theta_1 \tag{5.35a}$$

Two, one or none positive solutions exist, depending on whether the above expression is positive, zero or negative respectively. The discriminant is expanded as:

$$\mathfrak{D} = (\alpha\gamma)^2 - 2(\alpha\gamma)(\theta_2\beta + \theta_1\delta) + (\theta_2\beta - \theta_1\delta)^2 \tag{5.35b}$$

Setting $x = (\alpha\gamma)$, the above can be considered as a second-order equation:

[5] The characteristic equation takes the form $z^2 - Tz + D = 0$, where (T) is calculated as the trace and (D) as the determinant of the Jacobian. A necessary and sufficient condition for two stable roots is $/T/<1+D<2$, no matter if they are real or complex.

$$x^2 - 2(\theta_2\beta + \theta_1\delta)x + (\theta_2\beta - \theta_1\delta)^2 = 0 \tag{5.36}$$

It is trivial to show that the two roots (z_1, z_2) are given by:

$$z_1 = \left[\sqrt{\theta_i\beta} - \sqrt{\theta_1\delta}\right]^2 \tag{5.37a}$$

and

$$z_2 = \left[\sqrt{\theta_2\beta} + \sqrt{\theta_1\delta}\right]^2 \tag{5.37b}$$

The discriminant in (5.34) is positive if $\alpha\gamma < z_1$ or $\alpha\gamma > z_2$, and this leads to two real roots of (5.32). It is negative if $z_1 < \alpha\gamma < z_2$, and in this case no real root is obtained. Finally if $\alpha\gamma = z_1$ or $\alpha\gamma = z_2$, the discriminant is zero and the system has a single real equilibrium. Substituting (z_1, z_2) from (5.36, 5.37a), the three Propositions are readily obtained.

B. Wrong Lotka-Volterra Models

To demonstrate the inadequacies of autonomous Lotka-Volterra models with lagged interactions, the two types of models employed by Francisco (2009) are examined: one is used to study the dynamics of protests versus repression, and the other the dynamics of a civil war. Both of them render improbable behaviour, as shown in the examples below.

The Civil Conflict Model: Francisco (2009, Table 2.4, p. 17)

The interacting populations are assumed to represent protesters (*P*) and repression (*R*) in France for the period 1980–1995. Keeping the same notation, the system of interactions between adversaries is described as:

$$P_t = aP_{t-1} - gP_{t-1}R_{t-1} \tag{5.38a}$$

$$R_t = bR_{t-1} + hP_{t-1}R_{t-1} \tag{5.38b}$$

Estimated parameter values are reported as $a = 0.0161$, $b = 0.0738$, $g = 0.234 \times 10^{-3}$, $h = 0.2217 \times 10^{-5}$. For these values, it is easily found that the system has four equilibria (P^*, R^*) at (0,0), (417772,0), (0,−4205) and (417772,−4205). Of those, only the trivial equilibrium is meaningful and has two stable eigenvalues at (*a*, *b*). These are different from the pair {0.107,–0.00162) reported in Francisco (2009, p. 17) without any further details.

However, even the trivial equilibrium is problematic. Setting as initial conditions $P(0)=1000$ and $R(0)=1000$, the simulation gives $P(1)=-218$, $R(1)=76$, and the negative value is improbable. Moreover, the low modulus of eigenvalues suggests that the system converges too quickly to the trivial equilibrium, and thus is not suitable to capture long-term dynamics.

The Civil War Model: Francisco (2009, Table 5.2, p. 72)

Populations now are assumed to represent battle-deaths (R, S) for the two sides fighting in the US Civil War 1861–1868. Keeping the author's notation, the Lotka-Volterra model is described as:

$$R_t = \alpha S_{t-1} - mS_{t-1}R_{t-1} \tag{5.39a}$$

$$S_t = bR_{t-1} - nS_{t-1}R_{t-1} \tag{5.39b}$$

Estimated parameter values are reported as $a=0.1559$, $b=0.2629$, $m=0.349\times10^{-4}$, $n=0.868\times10^{-5}$.

Setting $R_t=R_{t\text{-}1}=R*$ and $S_t=S_{t\text{-}1}=S*$ and cross-substituting (5.38a) and (5.38b) we get

$$(1-ab)R^* = -(an+m)R^*S^* \tag{5.40a}$$

$$(1-ab)S^* = -(bm+n)R^*S^* \tag{5.40b}$$

The system (5.40a, 5.40b) has two equilibria. One of them is trivial:

$$R^* = 0,\ S^* = 0 \tag{5.41}$$

The non-trivial equilibrium is:

$$R^* = -\frac{1-ab}{bm+n},\ S^* = -\frac{1-ab}{an+m} \tag{5.42}$$

Clearly this is negative and meaningless for the estimated parameter values.

For the trivial equilibrium (5.40a), the Jacobian is given as

$$J = \begin{bmatrix} \dfrac{\partial R_t}{\partial R_{t-1}} & \dfrac{\partial R_t}{\partial S_{t-1}} \\ \dfrac{\partial S_t}{\partial R_{t-1}} & \dfrac{\partial S_t}{\partial S_{t-1}} \end{bmatrix} = \begin{bmatrix} 0 & a \\ b & 0 \end{bmatrix},\ \text{with eigenvalues} \pm\sqrt{ab} \tag{5.43}$$

The eigenvalues are calculated as (−0.20, +0.20) and are very different from the pair {0.1559, -0.50×10^{-4}) reported by Francisco (2009, p. 72) without details.

Setting as initial conditions $R(0) = 10{,}000$ and $S(0) = 10{,}000$ the simulation gives $R(1) = -1931$, $S(1) = 1761$, which are improbable. Again, eigenvalues have a very low modulus, and dynamics die away after a couple of periods. The model is clearly unable to portray the dynamics of a 7-year war.

The reason of both cases failing to realistically express conflict situations lies with the wrong structure of the model, as discussed in Sect. 5.3.6 of the present chapter.

References

Artelli M, Deckro R (2008) Modeling the Lanchester laws with system dynamics. J Def Modeling Simul 5(1):1–20. doi:10.1177/154851290800500101

Burbeck S, Raine W, Stark A (1978) The dynamics of riot growth: an epidemiological approach. J Math Sociology 6(1):1. doi:10.1080/0022250X.1978.9989878

Clausewitz C von (1976) On war. In: Howard M, Paret P (ed & transl). Princeton University Press, Princeton

Collier P, Hoeffler A (2007) Civil war. In: Sandler T, Hartley K (eds) Handbook of defense economics, vol 2. Elsevier, Amsterdam, pp 712–737

Din Q (2013) Dynamics of a discrete-time Lotka-Volterra model. Adv Differ Eq 2013:1–13, http://www.advancesindifferenceequations.com/content/2013/1/95

Epstein J (1997) Nonlinear dynamics, mathematical biology, and social science. Addison-Wesley, Boston, MA

Francisco R (1996) Coercion and protest: an empirical test in two democratic states. Am J Polit Sci 40(4):1179–1204

Francisco R (2009) Dynamics of conflict. Springer, New York

Guevara Che E (1961) Guerrilla warfare. https://armexploitednations.net/lib/-/guerrillawarfare.pdf

Helmbold RL (1965) A modification of Lanchester's equations. Oper Res 13(5):857–859, URL: http://www.jstor.org/stable/167735

Jaiswal NK (1987) Probabilistic analysis of combat models. Ann Oper Res 9:561–573

Jaiswal NK, Sangeeta Y, Gaur SC (1995) Stochastic analysis of combat models under different termination decision rules. Eur J Oper Res 83:530–546

Lanchester FW (1916) Aircraft in warfare: the dawn of the fourth aim. Constable, London

Lepingwell J (1987) The laws of combat? Lanchester reexamined. Int Secur 12(1):89–134, URL: http://www.jstor.org/stable/2538918

Linn III JB (2005) Force-using firms in the competitive equilibrium when public goods are absent. In: Eschenfelder (ed) Proceedings of Pennsylvania Economic Association, pp. 21–30. http://aux.edinboro.edu/pea/pub/Proceed2005/Contents2005.htm

Richardson L (1960) Arms and insecurity. Homewood, Pittsburgh (First edition 1919)

Simaan M, Cruz JB (1975) Formulation of Richardson's model of arms race from a differential game viewpoint. Rev Econ Stud 42(1):67–77

Taylor JG (1981) Mathematical analysis of nonlinear Helmbold-type equations of warfare. J Franklin Inst 311(4):243–265. doi:10.1016/0016-0032(81)90003-X

Tsebelis G, Sprague J (1989) Coercion and revolution: variations on a predator-prey model. Math Comput Model 12(4-5):547–559

Turchin P, Korotayev A (2006) Population dynamics and internal warfare: a reconsideration. Soc Evol History 5(2):121–158

Wallis PR (1968) Recent developments in Lanchester theory. Oper Res 19(2):191–195, URL: http://www.jstor.org/stable/3008650

Wei M, Chen G, Cruz JB Jr, Hayes L, Kruger M, Blasch E (2008) Game-theoretic modeling and control of military operations with partially emotional civilian players. Decis Support Syst 44:565–579

Zhang B, Zhang Z, Li Z, Tao Y (2007) Stability analysis of a two species model with transitions between population interactions. J Theor Biol 248:145–153

[illegible]

[illegible]

Chapter 6
Estimation of Conflict Dynamics

This chapter develops an econometric methodology to deal with the endogeneity problem in conflict models with contemporaneous interactions. Using the Greek Civil War time-series, a dynamic Lotka-Volterra model with external disturbances is estimated. A stable conflict trap is determined, explaining the prolongation of the civil war and its dire consequences for the country.

6.1 Estimation Issues

The dynamic model derived in Chap. 5 can be estimated by using time-series of battle losses and simple regression methods. As battle variables are co-determined, joint estimation of the system equations may entail the problem of 'simultaneous equation bias' (Greene 2000, ch. 15, p. 396). Possible non-stationarity of the variables in the system may also affect regression properties, as discussed below.

6.1.1 Model Specification

The problem of simultaneity in the estimation has already been mentioned in Chap. 1, and became clear from the theoretical foundations of the Lotka-Volterra model in Chap. 5. From Eqs. (5.28a, 5.28b) it is evident that the losses of one side do affect, and are at the same time affected by, the losses of the other party. In a multiple-equation least squares estimation framework it is inconsistent for either variable to appear as exogenously explaining the behavior of the other, while it is in fact co-determined. In this case, it is characterized by endogeneity and the

This chapter has been written in collaboration with Christos Axioglou.

N. Christodoulakis, *An Economic Analysis of Conflicts*,
DOI 10.1007/978-3-319-32261-2_6

disturbances in the independent variable are mixed up with the disturbances in the dependent variable. As a result, estimators will be biased and inconsistent, and the calculation of the conflict trap based on such estimated coefficients may be misleading.

Including fixed parameters for permanent disturbances as in (5.28a, 5.28b), the equations to be estimated are expressed by the following specification of the levels of time-series:

$$R_t = (1-\alpha)R_{t-1} + \beta R_{t-1}Q_t + \theta_1 + \xi_{1t} \tag{6.1a}$$

$$Q_t = (1-\gamma)Q_{t-1} + \delta Q_{t-1}R_t + \theta_2 + \xi_{2t} \tag{6.1b}$$

where (ξ_1, ξ_2) denote error terms that are normally distributed with a zero mean and possible nonzero correlation between each other.

6.1.2 Simultaneous Equation Bias

For the least squares estimator of the coefficients in (6.1a, 6.1b) to be unbiased and consistent, the error terms should be uncorrelated with the contemporaneous explanatory variables, i.e. $cov(R_{t-1}Q_t, \xi_{1t}) = cov(Q_{t-1}R_t, \xi_{2t}) = 0$. The expressions for the conditional expectations, implied by the model, are derived as:

$$E(\xi_{1t}, R_{t-1}Q_t | I_{t-1}) = R_{t-1} \cdot \frac{\delta Q_{t-1} var(\xi_{1t}) + cov(\xi_{1t}, \xi_{2t})}{1 - \delta\beta R_{t-1}Q_{t-1}} \tag{6.2a}$$

and

$$E(\xi_{2t}, Q_{t-1}R_t | I_{t-1}) = Q_{t-1} \cdot \frac{\beta R_{t-1} var(\xi_{2t}) + cov(\xi_{1t}, \xi_{2t})}{1 - \delta\beta R_{t-1}Q_{t-1}} \tag{6.2b}$$

where I_{t-1} stands for all information available in the system up to time $t-1$. As long as β or δ are nonzero and the error terms in Eqs. (6.1a) and (6.1b) exhibit nonzero cross-correlation, the application of least squares methods delivers biased and inconsistent estimates.

6.1.3 Stationarity

Conditions for stationarity of the battle variables are found in the relevant literature on random coefficient autoregressive (RCA) models (Nicholls and Quinn 2012) and locally stationary processes. (Dahlhaus 2012) provides a considerable amount of theoretical work on the characterization of such nonlinear processes. To link the present framework to this literature, Eqs. (6.1a, 6.1b) are re-written as

$$R_t = [(1-\alpha) + \beta Q_t]R_{t-1} + \theta_1 + \xi_{1t} \tag{6.3a}$$

$$Q_t = [(1-\gamma) + \delta R_t]Q_{t-1} + \theta_2 + \xi_{2t} \tag{6.3b}$$

In the above form it is obvious that each individual equation in the model describes a first order autoregressive process with stochastic/random autoregressive coefficients. Stationarity of (R) can be characterized according to whether the term $[(1-\alpha)+\beta Q_t]$ has a modulus greater than unity for some or all time periods; for (Q) the relevant term is $[(1-\gamma)+\delta R_t]$. This implies that Q and R share the same stationarity status, thus one needs to impose joint conditions on the evolution of both time sequences.

6.1.4 Estimation in First-Differences

A different specification is introduced by using first-differences, first in order to help reduce the effect of endogeneity as much as possible and, second, to obtain robust results in case variables are non-stationary in levels. Equations (6.1a, 6.1b) are now rewritten as:

$$\Delta R_t = -\alpha R_{t-1} + \beta R_{t-1}Q_t + \theta_1 + \xi_{1t} \tag{6.4a}$$

$$\Delta Q_t = -\gamma S_{t-1} + \delta Q_{t-1}R_t + \theta_2 + \xi_{2t} \tag{6.4b}$$

Defining all information available in the system up to time $t-1$ by the set $\{I_{t-1}\}$, the two conditions for endogeneity are specified as:

$$E(\xi_{1t}R_{t-1}Q_t|I_{t-1}) \neq 0 \Leftrightarrow E(\xi_{1t}Q_t|I_{t-1}) \neq 0 \tag{6.5a}$$

and

$$E(\xi_{2t}Q_{t-1}R_t|I_{t-1}) \neq 0 \Leftrightarrow E(\xi_{2t}R_t|I_{t-1}) \neq 0 \tag{6.5b}$$

We prove (6.5a) and the other part follows in an analogous way. First substitute Q_t with its equivalent expression from Eq. (6.1b) and assume that (ξ_1, ξ_2) are distributed i.i.d. over time and between each other. In this way we obtain:

$$E(\xi_{1t}Q_t|I_{t-1}) = \delta E(\xi_{1t}R_t|I_{t-1})Q_{t-1} \tag{6.6}$$

To continue, substitute R_t with its equivalent expression from Eq. (6.1a). By making the same assumptions as before we get:

$$\delta E(\xi_{1t}R_t|I_{t-1})Q_{t-1} = \beta\delta E(\xi_{1t}Q_t|I_{t-1})R_{t-1}Q_{t-1} + \delta E(\xi_{1t}^2|I_{t-1})Q_{t-1} \tag{6.7}$$

Combining this with the previous expression we have:

$$E(\xi_{1t}Q_t|I_{t-1}) = \beta\delta E(\xi_{1t}Q_t|I_{t-1})R_{t-1}Q_{t-1} + \delta E(\xi_{1t}^2|I_{t-1})Q_{t-1} \tag{6.8}$$

We now solve for $E(\xi_{1t}Q_t|I_{t-1})$ and obtain:

$$E(\xi_{1t}Q_t|I_{t-1}) = \frac{\delta\cdot E(\xi_{1t}^2|I_{t-1})Q_{t-1}}{(1-\beta\cdot\delta\cdot R_{t-1}Q_{t-1})} \tag{6.9}$$

The expectation term $E(\xi_{1t}^2|I_{t-1})$ in the numerator denotes the variance of ξ_1 and is positive. The RHS expression is clearly nonzero as long as δ is nonzero. This implies that the magnitude of the violation of the endogeneity assumption is:

$$E(\xi_{1t}R_{t-1}Q_t|I_{t-1}) = \frac{\delta E(\xi_{1t}^2|I_{t-1})R_{t-1}Q_{t-1}}{(1-\beta\delta R_{t-1}Q_{t-1})} \tag{6.10}$$

In the application to the Greek Civil War, estimates are obtained first by using OLS and then alternative methods that help correcting possible endogeneity biases.

6.2 APPLICATION: Estimating Conflict Dynamics in the Greek Civil War

As shown in Chap. 2, a structural shift regarding the pattern of casualties took place at the beginning of 1948. Therefore, dynamics should be separately examined in two phases: the first one covering the period from the breakout of the conflict in July 1946 to December 1947, while the second from January 1948 to September 1949 when the conflict was concluded with the defeat and formal surrender of DAG.

First the correlations of the variables appearing in (6.3a, 6.3b) are obtained as in Tables 6.1 and 6.2 for the two Phases of the conflict respectively. Results show that by using first-differences, correlation is low and therefore the problem of endogeneity is substantially reduced. Also, unit root tests similar to those conducted in Table 2.2 find both battle variables to be stationary.[1] The product terms in (6.4a, 6.4b) appear to be stationary in first differences, which makes the OLS estimation of their coefficient super-consistent.

[1]The only exception is with variable R in Phase I. Full details are available upon request.

Table 6.1 Correlations: phase I

Sample (adjusted): 1946M07 1947M12. Included observations: 18

Correlation coefficient (probability p)

	R	ΔR	R*Q(t−1)	Q	ΔQ	Q*R (t−1)
R	1	–	–	–	–	–
ΔR	0.407209 (p = 0.0935)	1	–	–	–	–
R*Q (t–1)	0.887633 (p = 0.000)	**0.018424 (p = 0.942)**	1	–	–	–
Q	0.493600 (p = 0.037)	**0.163591 (p = 0.516)**	0.491916 (p = 0.038)	1	–	–
ΔQ	0.284051 (p = 0.253)	0.482997 (p = 0.042)	**0.022758 (p = 0.928)**	0.687568 (p = 0.0016)	1	–
Q*R (t–1)	0.704489 (p = 0.001)	−0.253839 (p = 0.309)	0.881364 (p = 0.000)	0.625508 (p = 0.005)	**0.128012 (p = 0.612)**	1

Table 6.2 Correlations: phase II

Sample (adjusted): 1948M01 1949M09. Included observations: 21

Correlation coefficient (probability p)

	R	ΔR	R*Q(t − 1)	Q	ΔQ	Q*R (t− 1)
R	1	–	–	–	–	–
ΔR	0.689607 (p = 0.001)	1	–	–	–	–
R*Q (t−1)	0.527104 (p = 0.0141)	**0.047719 (p = 0.8373)**	1	–	–	–
Q	0.719129 (p = 0.0002)	0.638704 (p = 0.0018)	0.563011 (p = 0.0079)	1	–	–
ΔQ	0.614106 (p = 0.0031)	0.881538 (p = 0.000)	**−0.067794 (p = 0.7703)**	0.711535 (p = 0.0003)	1	–
Q*R (t−1)	0.600921 (p = 0.0040)	0.210573 (p = 0.3596)	0.806600 (p = 0.000)	0.783649 (p = 0.000)	**0.343798 (p = 0.1270)**	1

6.2.1 Single OLS Estimation

Below, the OLS estimation is adopted and results are displayed in Table 6.3. The main findings per phase of conflict are the following:

1. The exogenous adverse factors (θ_1, θ_2) rise substantially in Phase II for both sides, as the conflict becomes more aggressive. The rise, however, is more pronounced for DAG, reflecting a deterioration of guerrilla logistics and more impediments in tactical maneuvering after the closure of Greece's northern borders in 1949.

Table 6.3 Separate OLS estimates

	Phase I: 1946M07—1947M12		Phase II: 1948M01—1949M09	
Parameters	Estimate	p-value	Estimate	p-value
α	**1.066**	0.021	**1.157**	0.000
β	**6.797**	0.091	**1.268**	0.004
γ	**1.155**	0.001	**1.698**	0.000
δ	**0.927**	0.044	**0.665**	0.000
θ_1	**0.166**	0.064	**1.088**	0.001
θ_2	**0.068**	0.005	**0.237**	0.001
Equation statistics:	**(6.3a)**	**(6.3b)**	**(6.3a)**	**(6.3b)**
Nobs	18	18	21	21
Adj R-squared	0.267	0.479	0.647	0.715
S.E. of regression	0.239	0.039	0.433	0.127
F-statistic	4.101	8.817	19.292	26.108
Prob(F-statistic)	0.038	0.003	0.000	0.000
Akaike info criterion	0.127	−3.486	1.296	−1.152
Schwarz criterion	0.276	−3.338	1.446	−1.002
Hannan-Quinn criter	0.148	−3.466	1.329	−1.119
Durbin-Watson stat	1.679	1.812	1.616	1.564

2. Survival parameters (α, γ) are found to be nearly the same for both armies in Phase I, suggesting that the vulnerability of state troops was close to that of the insurgents. This is explained by the fact that state army losses included those of the Gendarmerie and other poorly-trained local militias that were fighting without any central coordination. The balance shifts in Phase II, during which survival rates hardly change for DAG but improve by nearly 50 % for GNA (from -1.157 to -1.698) owing to its increasing professionalization and the fact that all other forces were placed under its single operational command.
3. Firing effectiveness (β) of GNA remains superior to that of DAG (δ) in both phases of the war. However, both decline in Phase II as the armies get better organized and the conflict now involves larger-scale and deadlier battles rather than skirmishes.

Two reasons may explain these changes: First, the logistic support that DAG was receiving from abroad was sharply diminished after the Soviet Union withdrew its approval for the conflict and in April 1949 advised KKE leadership to end it.[2] Second, DAG became more vulnerable after Yugoslavia ceased to provide a safe backyard for retreating guerrillas and sealed its borders in July 1949.

Alternative estimations by using two- and three-stages Least Squares methods are reported in the Appendix of this chapter. Estimated coefficients are found to be

[2]According to his own testimony, the DAG leader was notified on 20/4/1949 that "Stalin put forward the case for retreating, for ending the armed struggle"; quoted in Rizospastis (2011, p. 449).

close enough with those obtained through OLS, but for one case that is discussed later. Hence, the consequences of ignoring possible correlation between (ξ_1, ξ_2) are not severe.

6.3 Evaluating the Conflict Trap

The findings of the econometric estimation imply that though GNA was virtually impossible to be defeated by the guerrillas, at the same time it was not capable of securing a quick victory over them either. DAG was broken down by inadequate replenishment, poor training and widespread inoperability of equipment. According to Marantzidis (2010, p. 97) no army can possibly last for long under such adverse factors. However, DAG was still mastering low-scale and mountainous engagements, and this made the conflict to take several turns before being terminated.[3]

To determine possible stalemates, two non-zero steady states are calculated for each set of estimated parameters; see Table 6.4. As the modulus of the characteristic roots exceeds unity, the higher equilibrium is found to be asymptotically unstable in both phases. In contrast, the lower equilibrium is asymptotically stable, suggesting the presence of conflict traps in both phases. Tellingly, both traps are found to be close to the historical average of battle deaths occurring in each phase. This implies that the conflict could have had been perpetuated around these levels for much longer, if exogenous factors had not been drastically altered in the meanwhile.

Table 6.4 Equilibria and stability

	Phase I: 46:07—47:12		Phase II: 48:01—49:09	
	Battle deaths (in '000s)		Battle deaths (in '000s)	
	Guerrillas	State army	Guerrillas	State army
High equilibrium	**0.618**	**0.117**	**1.622**	**0.383**
Eigenvalues	0.312	1.804	0.205	1.287
Stability	Two unstable roots		One stable, one unstable	
Low equilibrium	**0.314**	**0.079**	**1.481**	**0.333**
Eigenvalues	0.128	0.591	0.167	0.777
Stability	Two stable roots		Two stable roots	
Historical average	**0.309**	**0.084**	**1.364**	**0.370**

Notes: Variables are in '000s. An eigenvalue is said to be stable if and only if its modulus is below unity

[3]That outcomes are not necessarily determined by army numbers was indignantly expressed by a Government supporter who was sceptical about "the alleged mathematical assertions ... on so many more armies than bandits ... How then it happens that the former do not snatch the latter from the neck, to finish them off?"; daily Kathimerini 30/1/1049, reprinted in Rizospastis (2011, pp. 397–398).

A turning point occurred in the autumn of 1948, when GNA first cleared mountain Grammos but then failed to hold the neighbouring mountain Vitsi in northern Greece. The high toll of army casualties and the massive defections to DAG took Greek military leadership by surprise and demoralized its ranks.[4] According to Woodhouse (1976, pp. 144–145), the US mission was seriously considering to withdraw from Greece, and it was only after the visit of the State and Defence Secretaries to Athens in October 1948 that their engagement was reaffirmed. To regain control, the shaken Government appointed a hardliner veteran as Field Marshal and decided to plan new offensives against DAG strongholds.

Appendix

Alternative Estimations

To thoroughly examine the problem of endogeneity, a number of alternative estimates is obtained first and are then compared with the estimation results by using OLS estimation techniques. The following steps are taken:

Step 1*:* First, we need to take into account the presence of heteroskedasticity and/or contemporaneous correlation in the errors across equations. The seemingly unrelated regression (SUR) method, also known as the multivariate regression, or Zellner's method, is applied to estimate the parameters of the system.[5] Estimates are shown in Table 6.5.

Step 2*:* If some of the right-hand side variables are correlated with the error terms, but there is neither heteroskedasticity nor contemporaneous correlation in the residuals, the System Two-Stage Least Squares (S2SLS) estimator is appropriate. Estimation results are shown in Table 6.6.

Step 3*:* If the right-hand side variables are correlated with the error terms, and there is both heteroskedasticity and contemporaneous correlation in the residuals, a more appropriate method is the Three-stage least squares (3SLS) estimation.[6]

[4]Averof-Tositsas (2010, pp. 323–324) also claims that high ranking officials in the US were considering to opt out, while the Government was seriously contemplating defeat.

[5]EViews applies TSLS to the unweighted system, enforcing any cross-equation parameter restrictions. The estimates of the cross-equation covariance matrix are based upon parameter estimates of the unweighted system.

[6]EViews estimates STSLS by applying TSLS equation by equation to the unweighted system, enforcing any cross-equation parameter restrictions. If there are no cross-equation restrictions, the results will be identical to unweighted single-equation TSLS.

Table 6.5 Seemingly unrelated regressions estimation

	Phase I: 1946M07—1947M12		Phase II: 1948M01—1949M09	
Parameters	Estimate	p-value	Estimate	p-value
α	**1.302**	0.001	**1.179**	0.000
β	**8.980**	0.012	**1.311**	0.001
γ	**1.285**	0.000	**1.733**	0.000
δ	**1.147**	0.005	**0.683**	0.000
θ_1	**0.174**	0.029	**1.102**	0.000
θ_2	**0.072**	0.001	**0.241**	0.000
Equation	**(6.3a)**	**(6.3b)**	**(6.3a)**	**(6.3b)**
Nobs	36	36	42	42
R-squared	0.338	0.530	0.682	0.743
Adjusted R-squared	0.250	0.468	0.646	0.715
S.E. of regression	0.242	0.040	0.433	0.127
Durbin-Watson	1.451	1.605	1.559	1.506

Table 6.6 Two stage least squares estimation

	Phase I: 1946M07—1947M12		Phase II: 1948M01—1949M09	
Parameters	Estimate	p-value	Estimate	p-value
α	**2.374**	0.381	**1.033**	0.000
β	**20.032**	0.462	**0.469**	0.501
γ	**1.100**	0.002	**1.754**	0.000
δ	**0.756**	0.163	**0.714**	0.021
θ_1	**0.173**	0.148	**1.228**	0.001
θ_2	**0.069**	0.004	**0.234**	0.001
Additional Instruments:	$\{R_{t\text{-}2}S_{t\text{-}1}, S_{t\text{-}2}R_{t\text{-}1}\}$			
Equation	**(6.3a)**	**(6.3b)**	**(6.3a)**	**(6.3b)**
Adjusted R-squared	−0.33774	0.473353	0.559617	0.713302
S.E. of regression	0.323129	0.039477	0.483551	0.127801
Durbin-Watson	1.129185	1.88556	1.983945	1.497811
Instrument rank	5	5	5	5
J-statistic	1.3946	0.166116	1.463433	2.46322
Prob (J-statistic)	0.497928	0.920298	0.481082	0.291822

This amounts to a two-stage least squares version of the SUR method.[7] Results are shown in Table 6.7.

Step 4: Finally we compare the above estimates with the OLS outcome. Formal endogeneity is examined by employing the Durbin-Wu-Hausman test. The test

[7]These estimates are used to form an estimate of the full cross-equation covariance matrix which, in turn, is used to transform the equations to eliminate the cross-equation correlation. TSLS is applied to the transformed model.

Table 6.7 Three stage least squares estimation

	Phase I: 1946M07—1947M12		Phase II: 1948M01—1949M09	
Parameters	Estimate	p-value	Estimate	p-value
α	**2.550**	0.196	**1.023**	0.000
β	**21.789**	0.274	**0.463**	0.469
γ	**1.276**	0.000	**1.744**	0.000
δ	**0.944**	0.041	**0.711**	0.010
θ_1	**0.175**	0.097	**1.216**	0.000
θ_2	**0.077**	0.000	**0.233**	0.000
Equation	**(6.3a)**	**(6.3b)**	**(6.3a)**	**(6.3b)**
Included observations	36	36	42	42
Adjusted R-squared	−0.509	0.471	0.558	0.714
S.E. of regression	0.343	0.040	0.484	0.128
Durbin-Watson	1.136	1.632	2.001	1.515

stands for the existence of significant difference between the estimates obtained via OLS on each individual equation and those via Two Stage Least Squares.

Endogeneity Tests

Following the step highlighted in the previous Section, the system of Eqs. (6.4a, 6.4b) is estimated by applying the SUR method over the two phases. Using battle-deaths as the dependent variable, the following findings are established (results and further details available by the author)[8]:

(a) Serial correlation is not detected in the residuals; this implies no evidence of permanent effects by unobserved factors incorporated in (ξ_1, ξ_2).
(b) Additional experiments with alternative instrument sets deliver similar results. For example, by including $R_{t-1}S_{t-1}$ no perceptible difference is noticed.
(c) The Durbin-Wu-Hausman tests indicate statistically insignificant differences in all cases but for one marginally significant case.

By comparing estimates shown in Tables 6.5, 6.6 and 6.7, it seems that the coefficients are similar with those obtained by using OLS in Table 6.3.[9] Hence, the consequences of ignoring possible correlation between (ξ_1, ξ_2) are not severe. In all

[8]Estimation was also carried out for battle casualties and the results are similar to those reported for deaths. This is somehow embedded in the data as the figures for DAG fighters wounded in 1948–1949 were approximately set in Government records to be three times those of deaths; estimation details are available by the author.

[9]Unreported simulation results (available upon request), verify that simple least squares estimation performs fairly well in terms of bias and efficiency under different combinations of parameter values, including those reported in Tables 6.3, 6.5 and 6.6.

cases the explanatory power is satisfactory, and all coefficients but one are found to be statistically significant and correctly signed.

The only exception occurs regarding the firing effectiveness of GNA (β), whose estimate appears to be very sensitive to the choice of estimation method. By using the Seemingly Unrelated regression in Table 6.5, the estimates of (β) become significant and are closer to those obtained by simple OLS methods. However, the estimates obtained in Tables 6.6 and 6.7 are not statistically significant at the 10 % level.

References

Averof-Tositsas E (2010) By fire and axe: Greece 1946–1949 and the precedents. Estia Editions, Athens (In Greek)

Dahlhaus R (2012) Locally stationary processes. In: Time series analysis: methods and applications, vol 30. Elsevier, Amsterdam

Greene WH (2000) Econometric analysis. Prentice Hall, Upper Saddle River, NJ

Marantzidis N (2010) Democratic army of Greece (1946–1949). Alexandria Editions, Athens (in Greek)

Nicholls DF, Quinn BG (2012) Random coefficient autoregressive models: an introduction. Springer, New York

Rizospastis (2011) The three-year epic of the Democratic Army of Greece (1946–1949). Synchroni Epohi Editions, Athens (in Greek)

Woodhouse C (1976) The struggle for Greece: 1941–1949. Hurst, London

Chapter 7
A Model of Intertemporal Conflict

This chapter combines the conflict success function discussed in Chaps. 3 and 4 with the battle objective described in Chap. 5 into a unified framework. The two adversaries seek to optimize an intertemporal payoff function, subject to the army accumulation process that take into account battle losses and new recruits. The system is found to have unique saddle-path equilibrium and the effects of transient and permanent shocks are examined.

7.1 Introduction

A more promising way to formalize the dynamic characteristics of a conflict is to describe how battle tactics evolve over time. To that effect, an intertemporal framework, similar in principle to that of the Tullock model, is adopted. The two adversaries are assumed to decide how much battle effort to expend in order to *ex ante* maximize an intertemporal strategic objective. Dynamics are then obtained as the optimal tactics evolve in each period.

Thus far, the intertemporal problem has attracted little attention and it was only recently that Grossmann et al. (2011) developed a dynamic version of contest functions. However, the specifications are quite restrictive as the two competing parties are assumed to have the same effectiveness functions and reach the same steady-state values. No closed-form solution is obtained and the existence and stability of equilibrium is examined through linearization around the steady state.

7.2 An Intertemporal Conflict Model

The present framework extends the intertemporal model described by Grossmann et al. (2011) in three ways:

N. Christodoulakis, *An Economic Analysis of Conflicts*,
DOI 10.1007/978-3-319-32261-2_7

- First, fighting effectiveness is allowed to differ across the two rivals, since an asymmetric pattern is closer to actual situations. Asymmetric battle functions are also employed for describing the cost of engagement. In this way the model can be used to analyse conflicts between low-risk and high-risk adversaries.
- Second, the equilibrium levels of the two armies may differ as well.
- Third, uniqueness and stability of equilibrium is directly established without recurring to linearization.

7.2.1 Army Building

To avoid the criticism against deterministic continuity as in the Lanchester models, time is considered to be event-based and not calendar. In this way, time evolves by discrete events, separated from each other by an interval that does not necessarily remain fixed. Each event $t = 1,2,\ldots,$ describes a specific battle in which both armies expend a certain effort. Effort is measured by the amount of military equipment, ammunition, and/or human casualties. Whatever the means, battle casualties can be reasonably assumed to vary in some proportion to the effort. For this reason, the present framework assumes that effort is expressed by the level of losses suffered in the battlefield.

After the battle is concluded, each army is reorganized and prepares for the next fight. The fighting characteristics (A_j and m_j, $j = 1,2$) might have been altered because of the battle. For example, the winner may upgrade fighting effectiveness by capturing military equipment from the defeated, whose capabilities are thus diminished. Also, each army is adjusted by its losses and takes new recruits to augment the force. In the absence of recruits, the army would have declined at a non-zero rate due to fatigue, disease or other natural causes.

As a new battle event $t + 1$ is approaching, the two armies are adjusted according to the build-up processes:

$$r_{t+1} = (1 - \delta_1)r_t + x_{1t} - R_t \tag{7.1a}$$

$$q_{t+1} = (1 - \delta_2)q_t + x_{2t} - Q_t \tag{7.1b}$$

In the above expressions, (δ_1, δ_2) are the rates of decline between successive battles, (R_t, Q_t) denote battle-losses, while (x_1, x_2) are the new recruits for the guerrillas and state armies respectively. Initial conditions (r_0, q_0) are given. Recruitment involves selection and training; therefore it takes time to get completed and is based on information available at the end of previous event $(t-1)$.

Recruitments are subsequently treated as exogenous, to keep presentation as simple as possible. Note, however, that higher-order dynamics may be introduced by assuming that replenishments follow linear feedback rules on the previous period losses, e.g.

$$x_{1t} = \varphi_1 R_{t-1}, \quad x_{2t} = \varphi_2 Q_{t-1} \tag{7.2}$$

7.2.2 Battle Planning

The battle model draws many parallels from the consumer's intertemporal utility optimization problem in modern macroeconomics. Military resources accumulate like the stock of productive capital and are augmented by new recruits like new investments. A graphical representation of the process is shown in Fig. 7.1.

Battle outcomes are determined by the combination of forces that are lined up against each other, as well as by the effort and human sacrifice expended by the two sides in the field. Two assumptions are made:

1. forces shape the probabilities of winning as in the Tullock framework,
2. the battle payoff is an additively separable function of forces and effort.

Using a notation similar to that adopted in Chap. 4, the current-value payoffs are defined as follows:

$$\Pi_{1t} = p_{1t}(V_1 - L_1) - (D - L_1) + \beta_1 \cdot \Phi_1(R_t) \tag{7.3a}$$

$$\Pi_{2t} = p_{2t}(V_2 - L_2) + (\varepsilon D + L_2) + \beta_2 \cdot \Phi_2(Q_t) \tag{7.3b}$$

Note that the terms expressing the proportional cost of setting up the army (i.e. w_j, $j=1,2$) are now omitted since they now are implicitly taken into account by the accumulation constraints (7.1a, 7.1b) in the maximization of (7.3a, 7.3b).

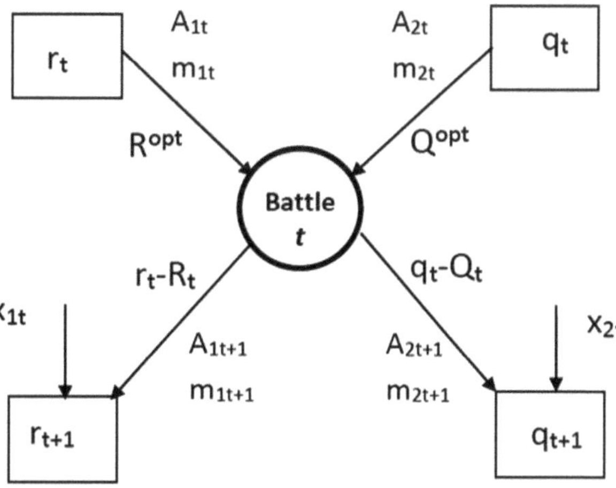

Fig. 7.1 The army formation process

Expressions $(\Phi_j, j = 1,2)$ are written as functions of the casualties suffered by the guerrillas and the state army respectively and denote the battle-effect brought about by expending a specific amount of effort in the field. They are increasing in the number of casualties, but the marginal effect diminishes as the toll accumulates, i.e. $\Phi_j^{'} > 0,\ \Phi_j^{''} < 0,\ j = 1, 2$. The rationale behind this assumption is that if a fighting unit suffers small losses, military equipment is still maintained and capabilities can be more easily recovered. However, if losses rise further, its fighting capability may be lost and the additional effect is limited. Explaining the increasing costs in warfare, Brauer and van Tuyll (2008, Ch. 4) argue that the acceptance of heavy losses is not rational when stakes fall short of resulting to the destruction of a whole system or the way of life. In such a case, additional battle losses would make potential new gains to be smaller and further costs to loom larger.

Parameters $(\beta_j,\ j = 1,2)$ express a trade-off between battle effort and resource preservation. Weighting is a matter of strategy and depends on the type of war that each side is determined to fight. Low values of (β_j) imply that preservation has priority over battle engagement and it corresponds to attrition warfare. High values imply over-engagement that may result in Pyrrhic outcomes, with a decimated army even in case of victory. The other terms are exactly as defined in Chap. 4.

The battle plan of each side $j = 1,2$ is to maximize a discounted flow of payoffs:

$$max\ \Omega_j = \sum_{t=1}^{\infty} \frac{1}{(1+\rho_j)^t} \cdot \Pi_{j,t} \tag{7.4}$$

In this framework, guerrillas maximize Ω_1 by planning to expend optimal effort (R^{opt}), while the state army maximizes Ω_2 by planning an optimal effort (Q^{opt}), and suffering the implied casualties.

Factors $1/(1+\rho_j)$ are weighting the importance of future battle outcomes versus current ones. In general, the rates take values in the range $-1 < \rho_1, \rho_2 < 1$. Positive rates indicate that a future battle is discounted, and this happens if a current one is likely to be more crucial for establishing a fighting position. As guerrillas in several cases need to overcome present difficulties by intense early fighting, and this is captured by assuming that $\rho_1 > \rho_2$. Negative rates lead to a weighting factor above unity and imply that future events are more critical than current developments.

7.2.3 The Information Set

The information on which battle effort decisions are based may not be fully available regarding the army accumulation variables. Each side is assumed to know its own forces, priorities, and fighting characteristics, but has only a limited guess about its opponent's size and fighting capabilities. The information sets available to each side before battle event *(t)* commences are defined as:

$$I_{1t} = \{r_t, A_{1t}, m_{1t}, q_t^e, A_{2t}^e, m_{2t}^e\} \tag{7.5a}$$

$$I_{2t} = \{q_t, A_{2t}, m_{2t}, r_t^e, A_{1t}^e, m_{1t}^e\} \tag{7.5b}$$

Since optimization takes place *ex ante,* decisions depend on the expectations that have been formed. As the battle evolves, decision variables are adjusted by the surprise effect and the true values are gradually revealed, thus $q_t^e \rightarrow q_t$ and $r_t^e \rightarrow r_t$. *Ex post,* effort variables may differ from *ex ante* optimal levels by a margin:

$$R_t = R_t^{opt} + \theta_{1t} \tag{7.6a}$$

$$Q_t = Q_t^{opt} + \theta_{2t} \tag{7.6b}$$

The terms (θ_j, $j = 1,2$) express unexpected battle shocks due either to the actual realization of variables being different from hypothesized levels, or other unpredictable developments. In terms of the information set, prior expectations are such that:

$$\varepsilon\{\theta_{jt} | I_{jt}\} = 0, \quad j = 1, 2. \tag{7.6c}$$

Notice that battle innovations are essentially similar to the exogenous shocks discussed in the Lotka-Volterra framework in Sect. 5.5.

The limitations on available information are taken into account in the definition of the payoff functions (7.3a, 7.3b). A key characteristic is that they include only the decision variable of the relevant side and not that of their opponent. The implication of this simplified structure is that the closed-loop reaction function that is obtained by applying optimal control methods is a Nash equilibrium, in the sense that a unilateral attempt of one side to deviate from it does not increase its payoff.

This makes a crucial difference from a situation where one assumes that payoffs include the adversary's effort as well. In that case, one side sees that a deviation by the opponent would affect its payoff and therefore it would have an incentive to deviate from optimal strategy as well. As shown by Simaan and Cruz (1975), such a structure requires a more complicated calculation to derive the Nash equilibrium solution.

In the present context of limited information, the optimal battle plan is derived by ignoring the opponent's current reaction in the payoff function due to the practical constraint that an inference on the other side's payoff, and thus an estimate of their current reaction function, cannot be obtained. Moreover, the opponent's current reaction is implicitly taken into account in the next period, since current losses reduce the size of the army and, therefore, affect the Conflict Success Function at battle event *(t + 1)*.

7.2.4 Optimal Battle Plans

Let's first proceed with the guerrillas' battle plan, and the government plan follows by analogy. According to Pontryagin's Maximum Principle, the Hamiltonian at event t is defined as:

$$\begin{aligned}\mathcal{H}_{1t} &= \Pi_{2t} + \lambda_{1,t+1} \cdot r_{t+1} = \\ &(V_1 - L_1)p_{1t} - (D - L_1) + \Phi_1(R_t) + \lambda_{1,t+1} \cdot [(1-\delta_1)r_t + x_{1t} - R_t]\end{aligned} \tag{7.7}$$

where $\lambda_{1,t+1}$ is the Lagrange multiplier for the army accumulation constraint. The first-order conditions for maximization are:

$$\frac{\partial \mathcal{H}_{1t}}{\partial R_t} = 0 \Rightarrow \lambda_{1,t+1} = \Phi_1'(R_t) \tag{7.8a}$$

and

$$(1+\rho_1) \cdot \lambda_{1,t} = \frac{\partial \mathcal{H}_{1t}}{\partial r_t} \Rightarrow$$

$$(1+\rho_1)\lambda_{1,t} = \frac{\partial p_{1t}}{\partial r_t} \cdot (V_1 - L_1) + (1-\delta_1)\lambda_{1,t+1} \tag{7.8b}$$

The second-order derivative gives:

$$\frac{\partial^2 \mathcal{H}_{1t}}{\partial R_t^2} = \Phi_1'' < 0 \tag{7.8c}$$

Expression (7.8a) denotes the shadow price of battle casualties. Lagging it one event to obtain the shadow price at event (t) and re-arranging (7.8b), the optimal battle effort is obtained as the implicit dynamic equation:

$$(1+\rho_1)\Phi_1'(R_{t-1}) - (1-\delta_1)\Phi_1'(R_t) = \Gamma_1\left(r_t, q_t^e\right) \tag{7.9}$$

Recalling the conflict success function p_1 from Chap. 4 [Eq. (4.3)] and taking derivatives, expression Γ_1 is defined for brevity as follows:

$$\Gamma_1(r,q) > (V_1 - L_1) \cdot \frac{\partial p_1}{\partial r} = \frac{m_1 \cdot (V_1 - L_1)}{r} \cdot \frac{A_1 A_2 r^{m_1} q^{m_2}}{[A_1 r^{m_1} + A_2 q^{m_2}]^2} \tag{7.10}$$

The function is well-defined in $(0, \infty)$ and positive everywhere. It is decreasing in both r and q.

7.2.5 *Battle-Effort Functions*

Closed-form solutions can be obtained if an effort function is specified. A wide class of functions with constant elasticity of marginal effect is defined as in Grossmann et al. (2011):

$$\Phi_1(R) = \frac{1}{1-n_1}R^{1-n_1} \tag{7.11a}$$

$$\Phi_1{}'(R) = R^{-n_1} > 0 \tag{7.11b}$$

$$\Phi_1{}''(R) = -n_1 R^{n_1-1} < 0 \tag{7.11c}$$

It is easy to show that

$$n_1 = -R \cdot \frac{\Phi_1{}''}{\Phi_1{}'} \tag{7.12}$$

It follows that parameter n_1 is the Arrow-Pratt measure of relative battle-risk aversion. Solving (7.9), the optimal battle plan is explicitly obtained by the expression:

$$(1-\delta_1)\cdot R_t^{-n_1} = (1+\rho_1)\cdot R_{t-1}^{-n_1} - \frac{1}{\beta_1}\cdot \Gamma_1\left(r_t, q_t^e\right) \tag{7.13}$$

7.3 Equilibrium and Dynamics

The steady states of battle losses and the army size are derived below, and then the stability properties and the reactions to shocks are examined.

7.3.1 *The Army Locus*

Examination takes place in the *(r, R)* space. Setting $R_t = R^*$ and $r_t = r_{t+1} = r^*$ in (7.1a), the accumulation process of the guerrilla army reaches a steady-state where:

$$r^* = (1-\delta_1)r^* + x_1 - R^* \Rightarrow R^* = x_1 - \delta_1 r^* \tag{7.14}$$

The locus is a downward-slopping line as shown in Fig. 7.2. If a transient shock perturbs the equilibrium by, for example, inflicting a higher level of losses, the RHS of (7.1a) implies that $\Delta r < 0$ and the system is driven back toward the locus, as indicated by the horizontal small arrows.

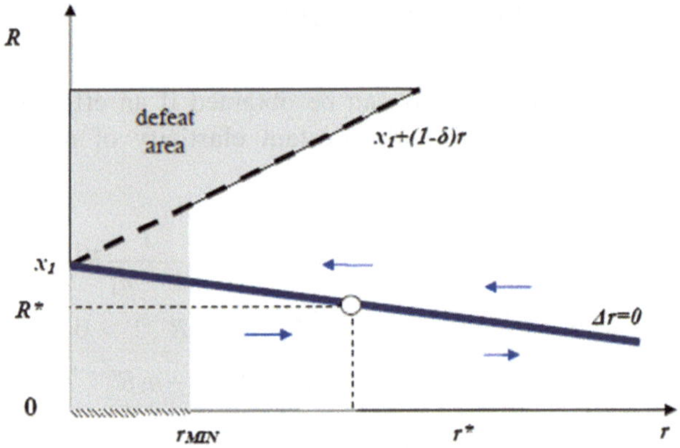

Fig. 7.2 Equilibrium locus for the guerrilla army

Extermination occurs if losses are too high and cannot be replenished by new recruitments. This happens when recruitments and battle survivors fall below the natural decline of the army, i.e.

$$x_{1t} + [r_t - R_t] < \delta_1 r_t \Rightarrow R_t > x_{1t} + (1 - \delta_1) r_t \tag{7.15}$$

For the guerrillas to enter the conflict in the first place, the threshold condition described in Chap. 4 must be satisfied, i.e. $r^* > r_{MIN}$. Applying the two constraints together, the shaded area in Fig. 7.2 indicates a combination of heavy losses and low army size that makes it impossible to continue the war. If the army falls within this area the conflict is terminated.

7.3.2 *The Battle Locus*

To find the locus of Eq. (7.9), we further set $q^e = q_t = q^*$, so that in equilibrium the guerrillas know exactly the size of the state army. After rearrangement the optimal battle plan should satisfy the steady-state equation:

$$(\rho_1 + \delta_1) \cdot \Phi_1'(R^*) = \Gamma_1(r^*, q^*) \tag{7.16}$$

For the family of battle-loss functions as in (7.11a), the equilibrium is explicitly obtained as:

$$R^* = [(\rho_1 + \delta_1)\beta_1]^{1/n_1} \cdot [\Gamma_1(r^*, q^*)]^{-1/n_1} \tag{7.17}$$

The locus in the (r, R) space is shown in Fig. 7.3. If a transient shock perturbs the equilibrium, for example by leading to a lower r^* level, the RHS of (7.9) implies

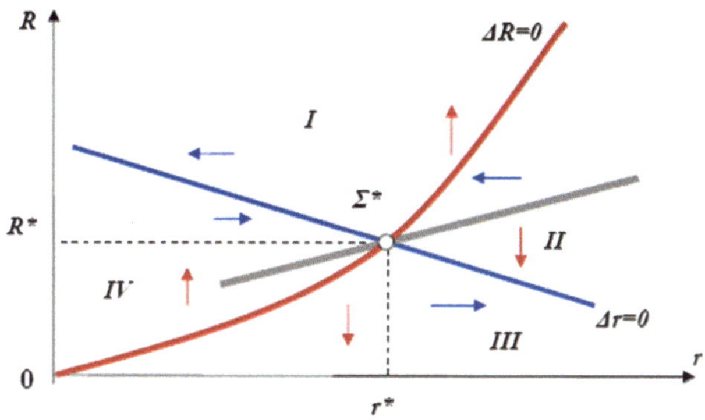

Fig. 7.3 Saddle-path equilibrium of the guerrilla army

that $\Delta R > 0$ and the system is driven further away from the locus, as indicated by the vertical small arrows.

7.3.3 *Stability and Response to Shocks*

If the two loci are combined as in Fig. 7.3, they result to an equilibrium (r^*, R^*). To examine the stability properties of the equilibrium, four quadrants are considered. It is clear that if a transient perturbation leads the system in the first or third quadrant, the dynamics are such that it is driven further away. It is only in the second and fourth quadrants that dynamics can be combined in such a way as to drive the system back to its initial equilibrium. The unique combinations form a saddle-path, on which the system is driven toward equilibrium. This saddle-path property allows for some interesting remarks on battle dynamics to be made.

Transient shocks will displace the system away from equilibrium and then a "jump" is needed to put it on the saddle-path and converge back to the initial steady state. Since the size of the army cannot change instantaneously, the "jump" will be implemented by a quick rise in battle losses. This type of adjustment leads to a high variability of battle losses as opposed to a low variability of the army size.

The effect of permanent shocks is analyzed in a similar way, as shown in Fig. 7.4. For example, suppose that the state army is permanently increasing its force from q^* to $q^{**} > q^*$. Since function $\Gamma_1(r, q)$ is decreasing in q but is also raised to a negative power, the battle locus will shift upwards, leading to a new equilibrium (R^{**}, r^{**}) that involves more losses and a smaller size of the guerrilla army. This implies that a larger battle with higher losses follows soon after the expansion of the state army.

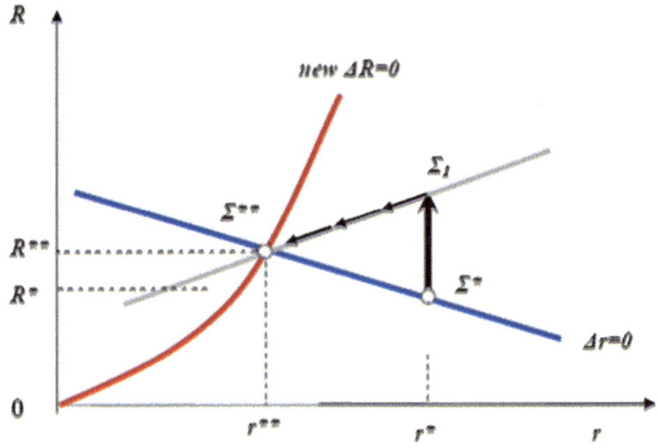

Fig. 7.4 Reaction to a permanent rise of the state army. *Notes:* After the initial jump to Σ_1, adjustment takes place along the saddle-path $\Sigma_1\Sigma^{**}$. The guerrilla army suffers more casualties and is reduced in size

7.3.4 The Dynamics of State Army

Working in an exactly similar way for the state army, the optimal battle effort is obtained as

$$(1-\delta_2)\Phi_2'(Q_t) = (1+\rho_2)\Phi_2'(Q_{t-1}) - \Gamma_2(r_t^e, q_t) \tag{7.18}$$

where it is set:

$$\Gamma_2(r,q) = \frac{m_2(V_2-L_2)}{q}\cdot\frac{A_1A_2r^{m_1}q^{m_2}}{[A_1r^{m_1}+A_2q^{m_2}]^2} \tag{7.19}$$

For specific loss functions for the two armies and with initial conditions *(r_0, q_0)*, Eqs. (7.9) and (7.18) can be solved in conjunction with (7.1a, 7.1b) to provide the *ex ante* evolution of battles over time.

7.4 The Battle Equation

An interesting relationship between battle efforts expended by guerrillas and the government is obtained. Rearranging (7.9) and (7.18), and dividing by parts we obtain

$$\frac{(1+\rho_1)\Phi_1'(R_{t-1}) - (1-\delta_1)\Phi_1'(R_t)}{(1+\rho_2)\Phi_2'(Q_{t-1}) - (1-\delta_2)\Phi_2'(Q_t)} = \frac{\Gamma_1(r_t, q_t^e)}{\Gamma_2(r_t^e, q_t)} \tag{7.20}$$

This expression relates the losses of both sides in each battle event with those of the previous battle event, the army size and the estimated guess of the size of the opponent's army. Some interesting relationships are examined.

7.4.1 The Steady State

In the steady state it is assumed that expectations on the opponents' armies reach the equilibrium values *(r^*, q^*)*. The steady-state implicit equation for the army sizes is obtained as

$$\frac{\Gamma_1(r^*, q^*)}{\Gamma_2(r^*, q^*)} \cdot \frac{r^*}{q^*} = \frac{m_1(V_1 - L_1)}{m_2(V_2 - L_2)} \tag{7.21a}$$

For brevity it is set

$$\Gamma = \frac{(\rho_2 + \delta_2)}{(\rho_1 + \delta_1)} \cdot \frac{m_1 \cdot (V_1 - L_1)}{m_2 \cdot (V_2 - L_2)} \tag{7.21b}$$

Substituting Γ_1, Γ_2 from (7.21a) into (7.20) and using (7.21b), the following equilibrium battle equation is finally obtained:

$$\frac{\Phi_1'(R^*)}{\Phi_2'(Q^*)} = \Gamma \cdot \frac{q^*}{r^*} \tag{7.22}$$

Equation (7.22) establishes a direct link between the losses of the two armies in case of a protracted conflict. The guerrillas sustain steady-state losses that positively depend on government losses, the relative "discriminatory power" (m_1/m_2), and the ratio of net rewards. They also depend on the relative size of the two armies, and guerrillas will suffer more losses if the size of the state army expands.

7.4.2 A Log-Linear Relationship

For the battle-effect functions specified in (7.11a), the battle equation becomes:

$$\frac{[R^*]^{-n_1}}{[Q^*]^{-n_2}} = \Gamma \cdot \frac{\beta_2}{\beta_1} \cdot \frac{q^*}{r^*} \tag{7.23}$$

Taking logs, the following log-linear expression of battle casualties is established:

$$ln\left[R^*\right] = \frac{n_2}{n_1} \cdot ln\left[Q^*\right] - \frac{1}{n_1} \cdot ln\left[\frac{q^*}{r^*}\right] - \frac{1}{n_1} \cdot ln\left[\Gamma \cdot \frac{\beta_2}{\beta_1}\right] \tag{7.24}$$

The slope of the log-casualties relationship is the ratio (n_2/n_1) of the relative battle-risk aversion of the state army and the guerrillas respectively. An empirical investigation is presented later in Sect. 7.5.

7.4.3 The Lanchester Analogy

The battle-ratio expression (7.24) is similar to the one derived in Sect. 5.1 for the Lanchester combat model with congestion effects. For strong congestion effects such that $a + b > 1$, setting $n = 1/(a + b-1) > 0$ and reorganizing (5.4a) the following ratio is obtained:

$$\frac{[dr]^{-n}}{[dq]^{-n}} = \left[\frac{h_1}{h_2}\right]^{-n} \cdot \frac{q}{r} \tag{7.25}$$

Given that there is no replenishment process in the Lanchester model, the change in the level of war resources occurs only through battle losses, thus $R = dr$ and $Q = dq$. The above ratio can be viewed as a special case of (7.23) by assuming that $n_1 = n_2 = n$, so that the two sides have similar risk attitudes toward battle losses. Both expressions imply that a power ratio of battle losses is proportional to the ratio of resources controlled by the two armies.

By comparing the two expressions, one may wish to describe combat parameters h_1 and h_2 of the Lanchester model in terms of parameters, preferences and fighting characteristics specified in the intertemporal problem.

7.5 Stylized Facts

The previous analysis can be used to explain some interesting patterns and characteristics of a conflict, such as volatility, escalation, and stalemates.

7.5.1 Conflict Patterns

Volatility is explained as a result of random shocks impinging upon battle calculations, as graphically shown in Fig. 7.4. The reaction of the two adversaries, as portrayed by the saddle-path adjustment, is to engage in new battles, implying abrupt changes and, therefore, high volatility in their casualties.

Escalation in the above framework is defined as an increase in battle-loss equilibria *(R^*, Q^*)*. Looking at expression (7.22) this may come from several factors that permanently affect the determinants of equilibrium. For example, a permanent rise of one army will lead to the escalation of the other, as examined before. If guerrillas suddenly become capable of raising their fighting discriminatory power (m_I), a permanent rise in the level of hostilities will follow, and so on.

A stalemate can be considered as a situation where de-escalation is not taking place and, at the same time, losses are sustainable by both sides. In current terminology this amounts to neither moving to a higher battle equilibrium that would be unattainable by one of the adversaries, nor finding a lower equilibrium that would gradually lead to the termination of conflict.

The role of discounting the future can also be examined. In discussing the reasons why negotiations fail to end a conflict, Skaperdas (2008) argues that this may happen when "a party that values the future highly could indeed take the chance of war instead of pursuing negotiation and compromise." In such a case, the benefits of an imminent termination of hostilities are dwarfed in comparison to the expected long-run profits after the opponents become permanently weakened or eliminated. This is captured by a negative rate of discount ($\rho_I < 0$). Given that the marginal battle-loss function is decreasing in (R^*), the reduction in the rate of discount raises (Γ) and thus shifts the battle locus downwards, as can be seen from Eqs. (7.21b) and (7.22). Dynamics are in the opposite direction as compared to the previous case: Guerrillas are not eager to undertake more battle losses and prefer to gradually accumulate more forces, thus undermining the prospects of a settlement.

7.5.2 The Log-Linear Pattern

An empirical investigation of the log-linear expression (7.24) is attempted below by using cross-section data from the Correlates of War dataset for the period 1900–2007. Aggregate data on total battle-deaths of the incumbent government (called "Side A") and the insurgent guerrillas ("Side B") are available for 115 civil wars worldwide.

Starting and end dates are also reported, and the duration of each conflict is calculated in months. In absence of time-series that would allow a proper investigation of battle equilibria in each particular country, an alternative is to consider aggregate conflict data as the sum of repetitive steady states, as in the static models described in Chap. 3.

Under this highly simplifying assumption, average monthly battle losses may be seen as an approximation of steady-states. Finally, to make data from different countries comparable, losses of each side engaged in the civil war are expressed per thousand of the country's population. Steady-states are set as the following indices:

Table 7.1 Analysis of cross-section civil war data 1900–2007

Statistics	Incumbent loss index (Side 'A'), in logs	Insurgent loss index (Side 'B'), in logs	Outcome
Observations	115	115	Balanced
Mean	−5.028	−4.14	–
Std. dev	2.31	2.01	–
Skewness	−0.091	−0.088	Skewed to the right
Normality Jarque-Berra test	0.164 (p = 0.92)	0.541 (p = 0.76)	Not rejected
Tests	*Method*	*Value*	–
Equal means	*t*-test	−3.10 (p = 0)	Rejected
Equal variances	F-test Siegel-Tuckey test	1.325 (p = 0.13) 0.966 (p = 0.33)	Not rejected
$n_1 = n_2$	Wald test	t-stat = 6.09 (p = 0) F-stat = 37.13 (p = 0)	Rejected

Notes: Cross-section data series are expressed as natural logarithms of the monthly averages of battle-deaths suffered by the incumbent side ('A') and the insurgent side ('B') per thousand of country's population. They cover 115 civil wars that took place worldwide during 1900–2007. Successive civil wars in the same country are grouped together.
Sources: Original data on civil war deaths and conflict duration are from the Correlates of War (COW) Project. Sarkees-Reid and Wayman (2010). *Resort to War: 1816–2007.* Washington, DC: CQ Press. The 1923 conflict in Italy was taken out of the sample as total losses were below 1000 and did not qualify as a civil war. Population data for 1900–1965 are the Maddison-Project (version 2013); for 1966–2007 from Worldbank-world development indicators.

$$Q^i = \left\{ \frac{battle\ deaths\ of\ side\ A\ in\ war\ (i)}{[pop\ (i)\ in'000] \cdot [months(i)]} \right\} \tag{7.26a}$$

$$R^i = \left\{ \frac{battle\ deaths\ of\ side\ B\ in\ war\ (i)}{[pop\ (i)\ in'000] \cdot [months(i)]} \right\} \tag{7.26b}$$

for state armies and guerrillas respectively and for cross-sections $i = 1,\ldots,115$. A number of statistical properties for the logarithmic series are summarized in Table 7.1 and discussed below:

1. The null hypothesis of a normal distribution of the series in logarithms cannot be rejected. Thus, the loss indices have a log-normal distribution.
2. The hypothesis of a common mean is rejected, but that of a common variance cannot be rejected at the 10 % level. The empirical probability density functions are depicted in Fig. 7.5.
3. No series is found to have a unit root.

Finally, the following cross-section equation is estimated (standard errors in brackets):

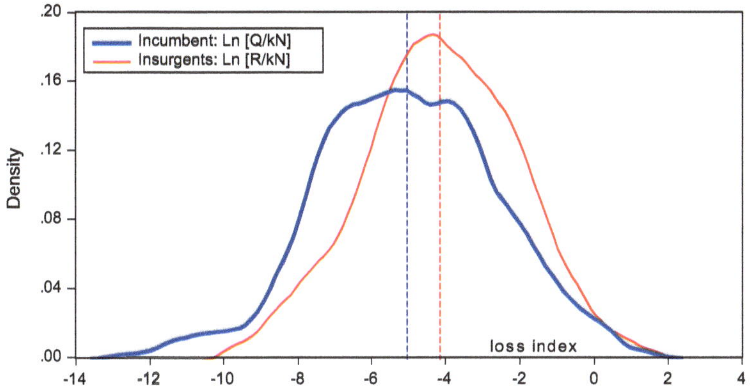

Fig. 7.5 Probability density function of the battle-loss index. *Notes:* Data are battle-deaths suffered by the guerrillas (R) and state army (Q) per thousand of country's population (N) and per month of conflict duration (k). In logs. *Sources* as Table 7.1

$$ln\left[R^i\right] = \underset{(0.044)^{***}}{0.7308}\, ln\left[Q^i\right] - \underset{(0.244)^{*}}{0.4658} \tag{7.27}$$

OLS, R2 adj = 0.705, Nobs = 115, SE = 1.094, F-stat = 273.73 (p = 0), DW = 1.92

The scatter diagram is shown in Fig. 7.6 with the 95 % confidence interval.

The slope is found to be statistically significant and less than unity. As shown in Table 7.1, the hypothesis of equal battle parameters $(n_2 = n_1)$ is heavily rejected by the relevant Wald test. Therefore, the relative battle risk-aversion parameter is lower for the state army. By a simple differentiation of (7.27), this finding suggests that a loss rate of $x\%$ for the incumbent army corresponds to a lower loss rate of around $0.73x\%$ for the insurgent army in average.[1]

It is interesting to see that the finding is re-affirmed even if one differentiates across continents. Table 7.2 shows that the slope is higher for civil wars in Europe, Middle East and Asia, and lower for Africa and Latin America. Nonetheless, it consistently remains below unity in all cases.

In the present context, the level of the incumbent's losses is found to be systematically lower than that of the insurgents, as implied by the mean-equality test shown in Table 7.1.[2] To inquire why this is happening, one can see that (7.11b) gives

[1]The coefficient is found to be very robust even if estimated from a sample including only European or non-European civil wars, before or after 1950 or any other time reference.

[2]The above finding evokes the discussion on the so-called Gibrat's Law in industrial economics, which employs log-series to test whether the growth rate of firms' activity in an industry is independent of the initial size of the firms. Empirical evidence is controversial; see, for example Sutton (1997).

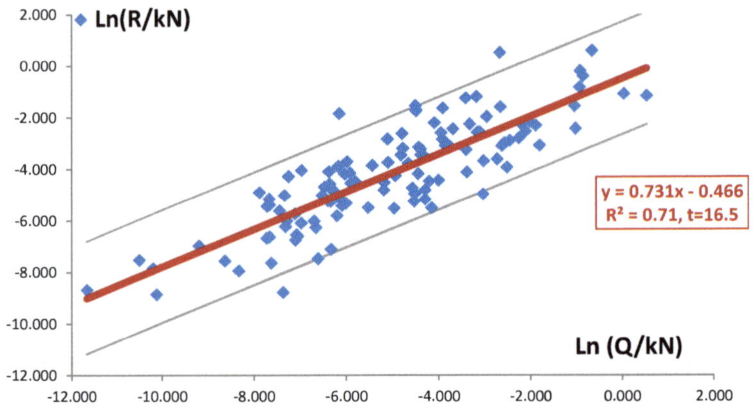

Fig. 7.6 The battle-loss index in 115 civil wars, 1900–2007. *Notes:* The parallel lines are at two standard errors from the main regression. *Sources* as in Table 7.1

Table 7.2 Battle elasticity by continent

Area	No. of conflicts	Slope	Std error (p value)	R2	S.E.R.
World	115	*0.731*	0.0441 (p = 0.0)	0.708	1.094
Europe	16	*0.787*	0.0708 (p = 0.0)	0.898	0.650
Middle East	17	*0.746*	0.1127 (p = 0.0)	0.744	0.942
Asia	38	*0.736*	0.0719 (p = 0.0)	0.744	1.058
Africa	22	*0.445*	0.1620 (p = 0.0125)	0.273	1.146
Latin America	22	*0.586*	0.1329 (p = 0.0003)	0.492	1.170

Note: Civil war data 1900–2007 from the Correlates of War (COW) Project. Sarkees Reid and Wayman (2010). *Resort to War: 1816–2007*. Washington, DC: CQ Press.

$$\Phi_1'(z) = z^{-n_1} < z^{-n_2} = \Phi_2'(z) \tag{7.28}$$

Therefore, for the same level of battle effort *(z)*, the marginal effect is larger for the state army than for the guerrillas. This makes the latter to be more cautious in further expanding their operations, and thus incur additional battle losses that are lower as a proportion to their current losses.

7.6 APPLICATION: The Dynamics of the Greek Civil War

Certain characteristics derived from the intertemporal battle framework are found to be in line with actual developments in the Greek Civil War as examined below.

7.6.1 Building Up the Guerrilla Army

In each period, the DAG force was increasing by either voluntary participation or forcible mobilization of populations in the occupied areas. At the same time, its size was curtailed by fighters surrendering to, or being captured by, the state army. Hence, net recruitments in each period are given by

$$RNET = RECR - RCAP - RSUR \tag{7.29}$$

Using simple OLS regression, an econometric estimate is obtained for the guerrilla army formation process, as described by (7.11a). The availability of new recruits was eventually severed after the government started implementing the policy of displacing rural populations to cut off supplies as discussed in Sect. 2.3.3. To capture this effect, a simple time trend is used. Battle casualties are denoted by *(R)* and the estimation is as follows:

$$\begin{aligned} DAG = {} & \\ & 3,239 + 0.908[DAG_{-1} + RNET_{-1}] + 0.901R_{t-1} - 0.580R_t - 144.04\,[time] \\ & (2,341) \quad (0.11)^{***} \qquad\qquad\qquad (0.21)^{***} \quad (0.20)^{***} (75.9)^{*} \end{aligned} \tag{7.30}$$

$Nobs = 31$, 1947:2-1949:08, $SEE = 2133$, $DW = 2.40$, $R2\ adj = 0.78$

Estimated coefficients are found to be statistically significant, correctly signed and in agreement with the build-up process described by Eq. (7.11a). The coefficient on current casualties is below unity, implying that only part of them is a permanent loss, while the rest eventually recuperate and re-join the army.

7.6.2 Endurance and Defeat

Using the graph in Fig. 7.4, the dynamic relationship between battle losses and the size of guerrilla army is examined. Separate graphs for the two Phases of the conflict are shown in Figs. 7.7 and 7.8 respectively.

All the way through Phase I, the guerrilla army gradually increases in size. Losses initially are low, but then a spike occurs around mid-1947. The spike is due to the increased mobilization of the state army that resulted in a number of special operations launched against guerrilla strongholds. The main operations were "*Ierax* (Falcon)" and "*Pelargos*" (Stork, 4/1947), "*Aetos*" (Eagle, 5/1947), "*Kyknos*" (Swan, 6/1947) and "*Korax*" (Crow, 5-8/1947). Though causing serious losses to the guerrillas, none of them proved to be decisive enough so as to defeat the insurgency and the guerrilla army continued to expand away from the defeat thresholds.

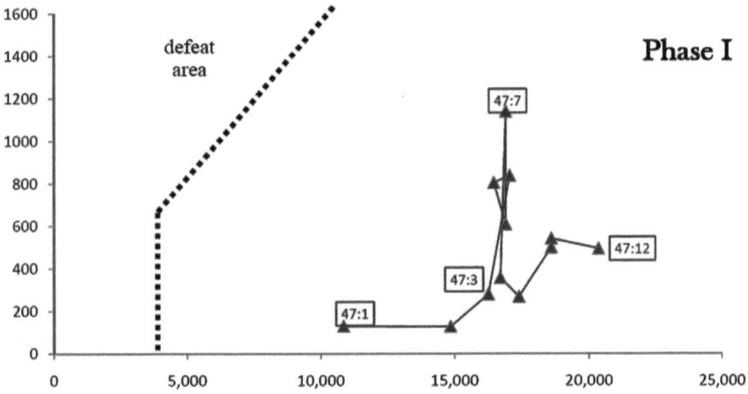

Fig. 7.7 The locus of the guerrilla army during Phase I. *Source*: Data for the period 1946:7–1947:12

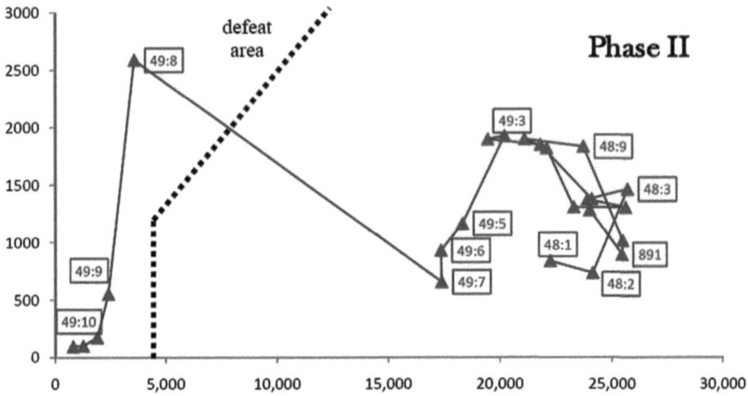

Fig. 7.8 The locus of the guerrilla army during Phase II. *Notes:* Data cover the period of the Greek Civil War 1948:1–1949:12. In the last months of the war, the locus had fallen within the *shaded area* shown in Fig. 7.2

In Phase II, the combinations of battle losses and army size seem to evolve around some equilibrium values until July 1949. (Derivation and properties of the equilibrium were given in Chap. 6). Finally, to resolve the impasse, an out-of-proportions escalation took place in the summer of 1949. The air force was intensively involved and aerial bombing reached unprecedented levels. These changes can be interpreted by a shift in the parameters expressing relevant fighting characteristics. It was due to this specific escalation, combined with the logistical collapse after all northern borders were sealed off, that DAG was terminally defeated and the conflict concluded. In Fig. 7.8, the system is shown to be catapulted in the defeat area from which it never recovered.

References

Brauer J, van Tuyll H (2008) Castles, battles, and bombs: how economics explains military history. University of Chicago Press, Chicago, US

Grossmann M, Lang M, Dietl H (2011) Transitional dynamics in a Tullock contest with a general cost function. BE J Theor Econ 11(1):1–26

Sarkees RM, Wayman F (2010) Resort to war: 1816–2007. CQ Press, Washington, DC, https://uk.sagepub.com/en-gb/eur/resort-to-war/book236426

Simaan M, Cruz JB (1975) Formulation of Richardson's model of arms race from a differential game viewpoint. Rev Econ Stud 42(1):67–77

Skaperdas S (2008) An economic approach to analyzing civil wars. Econ Gov 9:25–44. doi:10.1007/s10101-007-0043-2

Sutton J (1997) Gibrat's legacy. J Econ Lit 35(1):40–59

Chapter 8
The Economic Impact of Civil Wars: A Production Function Approach

The economic cost of a civil war arises from the destruction of production factors, the rise of uncertainty, and the misallocation of resources. An endogenous growth model shows how the re-allocation of resources from productive to unproductive activities suppresses the growth rate of the economy. The direct impact on output is evaluated by employing the growth-accounting framework on a production function with constant returns to scale. The Chapter describes how data can be used to provide estimates of the destruction of production factors and the loss of output in the Greek Civil War.

8.1 Economic Losses

A civil war brings about a multitude of economic losses, direct and indirect. Conflict suppresses growth, as it prevents normal economic activity, deters investment and physically destroys the factors of production. Another growth-impeding factor is the deterioration of productive infrastructure caused either by the destruction of physical infrastructure or the collapse of institutions. Many economic losses are not immediately observable or easily measured. Nevertheless, their direct and/or indirect effects on the economy's growth potential and the composition of output, and the population's prospects of well-being may be vast and long-lasting.

Considering the longer run, the fear of a civil war being likely to re-occur impedes growth through the alleviation of resources away from production and towards protection services. In this way, the quality and extent of production-supporting and human capital-enhancing public services are curtailed due to under-financing, as more government resources are channeled towards military procurement.

The neoclassical growth theory is usually employed to identify how the destruction of factors of production affects growth. On assessing the direct cost of civil war, an important assumption is the length of time that post-conflict consequences last.

N. Christodoulakis, *An Economic Analysis of Conflicts*,
DOI 10.1007/978-3-319-32261-2_8

Collier and Hoeffler (2007) assert that if conditions return to normal after the termination of hostilities, losses continue for another 21 years before GDP reaches the level that would have prevailed without the conflict. After peace is established, a growth rebound of around 1 % per year takes place and then gradually peters out as the economy returns to normal. Though such assertions are hard to apply universally, and the trajectory of a post-conflict economy is influenced by highly idiosyncratic events, the Collier-Hoeffler time frame is a useful benchmark for comparing losses across various civil wars.

8.2 Endogenous Growth in a Civil War

To overcome the lack of appropriate data in measuring the losses due to a civil war, a number of methods have been developed in order to determine the output loss through the detriment that war causes on the factors of production.

Murshed (2010, Ch. 2) considers a production system with capital accumulation and rent-seeking activities, modelled as in the Tullock framework. When the domestic conflict is aggravated, output and the marginal product of capital are reduced, leading to a decline of investment in the short and the long run. Although the curtailment of capital stock is an important consequence of conflict-related rent-seeking, the model leaves out all lasting effects that the civil war may have on savings and long-term growth. Such effects are better understood by employing endogenous growth theory, where the production function is extended to take all growth-inflicting factors into account.

Barro and Sala-i-Martin (1995, Ch. 4.4.2) examine two separate cases that are relevant in a conflict situation: in one the government uses public expenditure to exclusively finance productive infrastructure; in the other it devotes public revenues to secure the enforcement of property rights, and thus enhance the long-term rate of per capita output growth. Both modes of public revenue allocation are combined in a situation of rent-seeking activities developed by Economides et al. (2007). A model of atomized anarchy is described, in which households allocate their effort between productive and expropriation activities. The government chooses the optimal tax rate and the allocation of revenues between building infrastructure and law enforcement so as to maximize the rate of growth.

Both models need to be modified before applied in civil war context. In the case of a civil war, the use of violence against property is subservient to pursuing the overall strategy of the insurgents, and random acts of individual expropriation are usually forbidden as defamatory to the struggle. Thus the model of atomized anarchy is not suitable to analyse a civil war situation, since the threat to property rights comes from the collective violence produced by the conflict.

8.2.1 Infrastructure and Protection

In order to derive results that are analytically tractable, the allocation of revenues between infrastructure and law enforcement takes place as in Economides et al. (2007), while their effects are jointly accounted for, as in the framework described by Barro and Sala-i-Martin (1995). A closed economy is considered and assumed to be populated by N firms, each owning a private capital endowment (k_i), and using a linear technology to produce output $y_i = Ak_i$, $(i = 1,2,\ldots,N)$, with (A) a technology parameter. Aggregate output is denoted by (Y).

In each period of time $(t, t = 1,2,\ldots)$ the capital stock depreciates at a rate (δ). Its rate of return is such that the after-tax marginal product is equal to the rental cost of private capital. Households have a subjective rate of discount equal to (ρ), and maximize an intertemporal utility function of current consumption with a constant relative risk aversion (φ). More details are given in the Appendix.

The government imposes a constant uniform taxation (τ) on produced output in order to finance public spending (G) without accumulating public debt, that is $G = \tau Y$. A part $D = \omega G$, $(0 < \omega < 1)$ of public expenditure goes to security and defense spending and the rest $S = (1 - \omega)\, G$ to productive infrastructure. As will be shown below, the fraction (ω) tends to increase when a conflict takes place, and is reduced in peace time.

Security services are used to safeguard property rights, so that private producers retain their capital stock and have a motivation to invest and raise production. The extent of enforcing property rights is captured by a probability function $\pi(D)$, with $0 \leq \pi \leq 1, \pi' > 0, \pi'' < 0$. Enforcement rises as policing (D) increases, but with diminishing marginal effect. A suitable enforcement function may take the form

$$\pi(D) = 1 - e^{-\lambda D} \tag{8.1}$$

Parameter (λ) denotes the effectiveness of public services on enforcing property rights. Obviously, $\pi(0) = 0$ so that in the absence of security spending there is no protection. With either (λ) or D sufficiently large, we have $\pi \to 1$ and full protection is achieved.

Alternatively, the probability of the output being expropriated is $1 - \pi = e^{-\lambda D}$, which is inversely related to the effectiveness of security. The eruption of a civil war increases the threat of output expropriation and this can be modeled as a step reduction in the enforcement parameter (λ).

Public infrastructure is characterized by congestion effects and exerts a positive impact on private production. One way to model congestion is by taking the positive impact to be proportional to the output produced. A suitable function is given by

$$h(S) = S^{\beta} \tag{8.2}$$

where (β) denotes the output elasticity to public infrastructure and $0 < \beta < 1$ ensures decreasing returns to scale, i.e. $h' > 0$ and $h'' < 0$. Substituting $G/Y = \tau$, $D = \omega\tau$, $S = (1-\omega)\tau$, and omitting subscript (i) for simplicity, private firms are expecting to produce and retain an output that is given by:

$$y = \pi(\omega\tau) \cdot Ak \cdot h[(1-\omega)\tau] \tag{8.3}$$

Following Barro and Sala-i-Martin (1995), it can be shown that the growth rate of per capita output is given by:

$$g = \left[\frac{1 + \Lambda(\tau) - \delta}{1 + \rho}\right]^{1/\varphi} - 1 \tag{8.4}$$

where the following expression is used for brevity:

$$\Lambda(\tau) \equiv (1-\tau) \cdot A \cdot \pi(\omega\tau) \cdot h[(1-\omega)\tau] \tag{8.5}$$

Derivation and further details are given in the Appendix.

8.2.2 Growth-Enhancing Policies

In order to maximize the growth rate, the government has to make two decisions: first to choose the optimal tax rate (τ), and then to allocate a share (ω) of revenues to security and the rest $(1-\omega)$ to productive infrastructure. Given that the above growth rate is an increasing function of expression (Λ), the government's decisions are described by setting

$$\frac{\partial \Lambda}{\partial \tau} = -A\pi h + (1-\tau) \cdot A \cdot \left[\omega\pi' h + (1-\omega)\pi h'\right] = 0 \tag{8.6a}$$

and

$$\frac{\partial \Lambda}{\partial \omega} = (1-\tau) \cdot A \cdot \tau \cdot \left[\pi' h - \pi h'\right] = 0 \tag{8.6b}$$

Rearranging (8.6a) and dividing by (πh) we obtain the growth-maximizing condition:

$$\frac{1}{1-\tau} = \omega\frac{\pi'}{\pi} + (1-\omega)\frac{h'}{h} \tag{8.7a}$$

Also from (8.6b):

$$\frac{\pi'}{\pi} = \frac{h'}{h} \tag{8.7b}$$

Equation (8.7a) is the efficiency condition for allocating public revenues between security and productive services. Substituting into (8.7b), the optimal taxation (τ^*) and optimal security share (ω^*) are easily obtained by the following equations:

$$(1+\beta)\cdot\tau = \beta + \frac{1}{\lambda}\cdot ln[1+\lambda(1-\tau)] \tag{8.8a}$$

$$\omega = 1 - \left(\frac{1}{\tau} - 1\right)\beta \tag{8.8b}$$

where *ln(.)* stands for natural logarithms. The following Propositions hold:

Proposition 1 The optimal tax rate is a decreasing function of enforcement effectiveness *(λ)*, i.e. $\partial\tau^*/\partial\lambda < 0$.

Proposition 2 The optimal budget share of enforcement policies is a decreasing function of enforcement effectiveness *(λ)*, i.e. $\partial\omega^*/\partial\lambda < 0$.

Proposition 3 The level of public resources allocated to enforcement policies is a decreasing function of enforcement effectiveness *(λ)*, i.e. $\partial\left(\omega^*\tau^*\right)/\partial\lambda < 0$.

Proofs are given in the Appendix.

8.2.3 *The Effect of a Civil War*

Applying comparative statics, the above properties may be utilized to illuminate a number of interesting facts that are observed during civil wars.

1. *Spreading of violence:* As the conflict intensifies, the enforcement of property rights weakens and this can be expressed by a step reduction in the effectiveness parameter *(λ)*. This implies that the government has to increase distortionary taxation in order to raise more revenues and finance a higher level of security. A graphical demonstration is given in Fig. 8.1.
2. *Militarization of production:* With a decline in security effectiveness, the government increases the allocation of resources to defense and security (higher ω and higher $D = \omega\tau$). Subsequently this increases the role and influence of the military in ensuring and protecting production and the distribution of output.

   The implication of the above remarks is that as civil war intensifies and violence spreads, the economy suffers a double cost: first from the rise in distortions brought about by a higher tax rate and, second, from the curtailment of productive infrastructure. The long-term growth rate is cut as shown in Fig. 8.2.

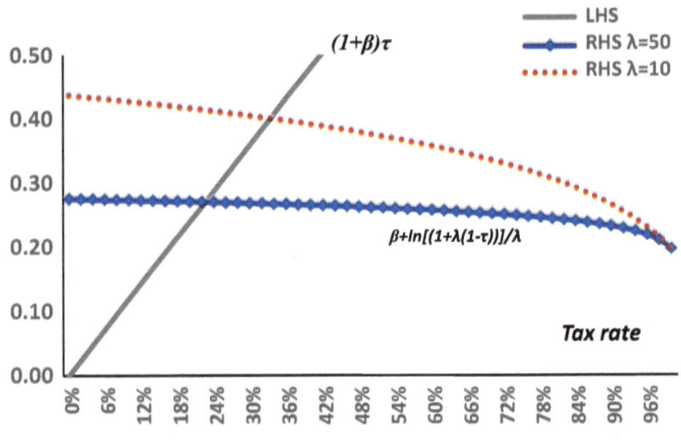

Fig. 8.1 Optimal tax rate for different levels of enforcement effectiveness. *Notes:* The vertical axis measures the LHS and the RHS of Eq. (8.8a). Parameters are set as $\lambda = 50$, $\beta = 0.2$. For illustration, weaker enforcement (*dotted line*) described by setting $\lambda = 10$

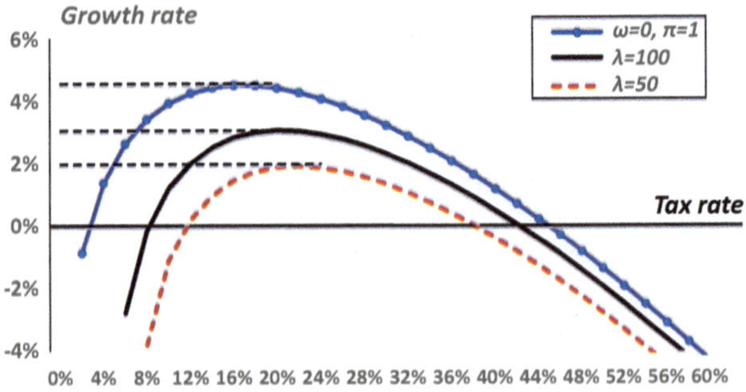

Fig. 8.2 Growth-maximizing tax rates and property rights enforcement. *Notes:* Parameters are set as $\lambda = 100, \beta = 0.2, A = 0.32, \delta = 10\,\%, \theta = 5\,\%, \sigma = 0.80$. Weaker enforcement (*dotted line*) at $\lambda = 50$. Full protection of property rights (*marked line*) is obtained as $\lambda \to \infty (\omega = 0)$

3. *Emergency measures:* The above analysis assumed that production is linear in the enforcement of property rights. In practice, however, production will be undertaken only if the probability of the firm retaining its output surpasses a certain threshold, $\pi \geq \pi_0$. From (8.1) it is clear that this requires a minimum of security effectiveness, so that

$$\lambda \geq \lambda_0 := -\frac{1}{\omega\tau}\ln(1-\pi_0) > 0 \tag{8.9}$$

The problem with satisfying the above condition is that both the protection budget share and the tax rate are constrained in the interval *(0,1)*, so that the threshold cannot fall below a certain level. In case that effectiveness falls too low, the only way to restore condition (8.9) is to raise *(D)* exogenously, for example, through aid from abroad and/or by a foreign intervention.

8.3 Empirical Evidence

The detrimental effect that a civil war, or political instability in general, exert on economic growth has been empirically investigated in several studies. However, a caveat is due before such empirical evidence is presented. Despite its theoretical appeal, the production function approach described above faces serious problems when applied in practice. In most cases, it is unlikely that adequate data on the destruction of production factors or the downgrading of human capital formation are available for estimation. Nor is it possible to gather any data—let alone reliable ones—that measure the extent and effectiveness of enforcing property rights so that their effect on the growth rate can be quantitatively assessed.

The most common approach is to obtain estimates of the adverse effect that a conflict inflicts upon the growth rate of an economy by pooling data from different countries with and without a civil war. In such studies, an implicit assumption is made that all pooled countries share common economic and institutional characteristics and that the incidence of a civil war impacts on them in a similar pattern. To mitigate those assumptions, idiosyncratic country effects are taken into account in various estimations. With this caveat in mind, growth-reducing effects caused by a civil war are established in a number of studies.

For example, Alesina and Perotti (1996) estimated a cross-section of 71 countries over the period 1960–1985 and obtained that political instability leads to the curtailment of physical investment. They also established that income inequality is a channel that produces political instability by increasing social discontent and fueling unrest and various acts of political violence.

Barro and Sala-i-Martin (1995, Ch. 12) use data from 87 countries for the period 1965–1975 and 97 countries for the period 1975–1985, and examine how growth is affected by a variety of socio-economic factors. An index of rebellions and political violence is meant to express political instability and the threat to property rights that adversely affects investment and production activities. An inverse relation of political instability and the growth rate is established by showing that a rise in the political instability index by one-standard deviation lowers the growth rate by 0.40 % per year.

Imai and Weinstein (2000) describe that the economic cost of a civil war stems mainly from the deterrence of private investment due to the increased risk of property expropriation. Another channel is the strain that a conflict puts on fiscal deficits, leading to further instability and reduction of growth. Using data from 104 countries over four decades from the 1960s to the 1990s, they show that the decade-average growth rate of per capita income is reduced by 0.20 % annually as a result of a widespread civil war.

A study that is more specific on civil wars is conducted by Murdoch and Sandler (2002), who employ a neoclassical growth model with human capital to examine civil-war effect on per capita output of the country involved, as well as the collateral damage on neighbouring economies. Using empirical data they find strong econometric evidence that the incidence of a civil war adversely affects the growth of per capita income of the country involved in the short-run period of 5 years.

The adverse effect is less pronounced in the long-run of 20–25 years and the reason is that after a civil war terminates, the low-income and low-accumulation country enters a process of convergence with other economies. However, the fact that an adverse effect is still detected in the long-run is an indication that even if a civil war terminates it leaves a legacy of growth-hurting factors in the country. Similar effects are found for the neighbouring countries.

Furthermore, an increase in military and security spending may impose other costs on the economy, as it enhances secrecy and corruption in decision-making. Gupta et al. (2001) use annual data for up to 120 countries in the period 1985–1998 to establish a positive relationship between corruption and military spending measured as a share of GDP. Then a conflict dummy was introduced, depending on whether or not the country was involved in external or internal armed conflict. In the military spending equations the conflict-country dummy was found to affect military spending positively, as expected, and to be statistically significant.

Another approach that indirectly assesses the economic impact of a civil war is by comparing the specific country or region engaged in the conflict with similar areas that stay away from hostilities. This method is analysed in the next Chap. 9.

8.4 Measurable Direct Costs

The examination of the growth rate is very useful for understanding the channels through which output destruction occurs in a civil war, but is mired with difficulties in finding the appropriate data for empirical estimation in specific economies. Variables such as public security and productive infrastructure are difficult to quantify, especially for less developed economies during war periods. Human capital effects are also difficult to quantify, as this requires a distinction of human losses per educational category or labor division.

Nonetheless, some types of production factors are available in time-series, and sometimes the damage they experience during a conflict is also recorded. As before, we assume that there are *(N)* productive industrial units in the economy, and all of them are identically invested with a capital stock per factory equal to *(k)*. Thus, total productive capital is given by $K = kN$. There is also an agricultural sector, which is using livestock *(Z)*. Employment is *(L)*. Assuming constant returns to scale across the economy, output in constant prices *(Y)* is given by:

$$Y = A(kN)^{\eta} Z^{\varepsilon} L^{1-\varepsilon-\eta} \tag{8.10}$$

Parameter (A) is a technology factor, while $(\eta,\ \varepsilon)$ are the respective stock elasticities of output. The direct economic cost of a civil war is estimated by calculating the destruction of production factors and the resulting suppression of output brought about by the conflict. Let dN, dZ and dL denote the destruction of factories, livestock and employment respectively. Taking log-derivatives, the growth accounting identity is obtained as:

$$\frac{dY}{Y} = \frac{dA}{A} + \eta\frac{dK}{K} + \varepsilon\frac{dZ}{Z} + (1 - \eta - \varepsilon)\frac{dL}{L} \tag{8.11}$$

If the conflict-related destruction of labour force, production capital, and livestock can be measured, the above expression gives the output loss. The first term in the RHS denotes the fall in the growth rate that is due to a technology breakdown and may result from a loss of expertise, destruction of high-tech units or other reasons that cause total factor productivity to decline. The other terms express the extent to which particular production factors have been destroyed.

8.5 APPLICATION: Estimating the Direct Cost of the Greek Civil War

In the case of the Greek Civil War, capital stock data are not available. Besides, there is no detailed account on human capital losses on the basis of the educational and professional qualifications of those perished or exiled. However, there are quite reliable data regarding aggregate human losses and also evidence on the expropriation of livestock, as well as on industrial destruction caused by sabotage actions during the conflict.

8.5.1 Assumptions

Under some simplifying assumptions on the patterns of production and technology, an estimate can be obtained for GDP losses due to the civil war. The assumptions are described below:

(a) The capital stock *(k)* per production unit is assumed to having remained the same before and after WW2. If each firm produces an output *(v)*, total industry output is $V = vN$ and its growth rate is given by $dV/V = dN/N$. Therefore, the loss of the capital stock can be approximated by the number of factories that were destroyed during Civil War hostilities or sabotaged as a pre-emptive strike to cut critical supplies.
(b) Livestock *(Z)* was heavily reduced during the conflict, for two reasons: first because guerrillas were frequently raiding villages to confiscate agricultural produce and animals for food supplies; second because large herds were abandoned after the evacuation policy was enforced in the countryside. After the conflict ended, most of such losses were recorded by their owners in claims of compensation, thus data are available to measure the loss in livestock.

(c) As no distinction can be made regarding the skill differentiation of human losses, all of them are treated as a reduction of the uniform labor force *(L)*, measured by the size of working-age population. During the Civil War, the active population was reduced by the number of battle-deaths, severe battle-casualties, and by those sentenced to death or convicted to long-term internment. After the termination of fighting, it was further reduced by the fugitive guerrillas and their families who fled Greece to avoid further persecution.
(d) For simplicity it is assumed that technology *(A)* remained unaffected by the conflict (i.e. $dA = 0$).

8.5.2 Growth Accounting

Let (ψ) denote the proportional loss of output, i.e. the drop from what could have prevailed in the absence of the Civil War. Taking into account the above assumptions and using subscript CW to denote the Civil War period, the proportional drop in output is given by the formula:

$$\psi = \frac{dY}{Y} = \eta \frac{d(N_{CW})}{(N_{CW})} + \varepsilon \frac{dZ_{CW}}{Z_{CW}} + (1 - \eta - \varepsilon)\frac{dL_{CW}}{L_{CW}} \tag{8.12}$$

Factor destructions are calculated as proportions of the initial stocks, as described below:

1. Employment:
   Let (P_t) denote the active population in period t. According to Table 8.1, human losses in the Civil War amounted to $dP_{CW} = -236{,}787$ persons. To obtain the losses in employment, it is further assumed that the proportion of unemployed (u) among those perished or expatriated was the same as in the total active population. The nearest Census of active population took place in 1951 and reflected the aforementioned reductions; therefore the loss in employment due to the Civil War is calculated by the adjusted formula:

$$\frac{dL_{CW}}{L_{CW}} = \frac{(1-u)\cdot dP_{CW}}{(1-u)\cdot P_{CW}} = \frac{dP_{CW}}{P_{CW}} = \frac{dP_{CW}}{P_{1951} + dP_{CW}} \tag{8.13}$$

   Using data from Table 8.1, the aggregate loss of employment is found to be −7.80 % of total labour force.
2. Livestock:
   From Table 8.1, we can see that $dZ_{CW} = -1{,}480{,}669$ animals were killed during the hostilities. The nearest data available for total livestock are from the agricultural Census of 1951, thus a similar adjustment applies as in (8.14) and the destruction is calculated to be −11.52 % of total livestock. This probably underestimates the loss due to the Civil War, as the calculation leaves out the

Table 8.1 Human toll and GDP losses due to the Civil War

	Factors of production	Actual losses	Terminationin June 1949
Human capital			
1	Total battle deaths	43,452	38,328
2	Seriously wounded (b = 2a)	86,904	76,656
3	Ex-patriated by KKE	55,881	0
4	Sentenced to death	4832	4664
5	Sentenced to more than 10 years	45,718	0
6	Total losses in human capital, (1) to (5)	236,787	119,648
7	Active population 1951	2,800,413	
8	As % of active population adjusted by the losses	−7.80 %	−3.94 %
Industry			
9	Industry, growth rate 1934–1938, annual average	7.3 %	
10	New factories 1934–1938, annual average	124	
11	Factories destroyed during the GCW	241	
12	As % of industrial units	−15.76 %	
Livestock			
13	Destruction during 1946–1949 (animals)	1,480,669	
14	Livestock in 1950 (animals)	11,374,600	
15	As % of livestock adjusted by the losses	−11.52 %	
Growth accounting			
16	Non-agricultural capital remuneration, % GDP 1954	0.360	
17	Agricultural income, as % GDP 1954	0.315	
18	Estimated growth rate loss during GCW	−2.84 %	−2.43 %
19	Total GDP loss discounted at 5 %	128.55 %	89.70 %

Notes and data sources
(1) The sum of data series SKLD+RKLD. Averof-Tositsas (2010, pp 384–385) claims that battle deaths were 36,839 guerrillas and 14,356 from GNA. Thus present calculation may be conservative
(2) Seriously wounded are estimated as twice the number of deaths
(3)Papathanasiou (2002, p 147). Of those 17,352 were children, but here are accounted as active population as most of them reached working age within a few years
(4, 5) Michiotis (2007, pp 235–239). (7) ESYE (1951, Table 1, pp 2–9)
(9, 10) ESYE (1939, Table B1, pp 123–124). (11, 13) As quoted in Rizospastis (2011, p 564)
(14) ESYE (1958, Table IX, p XXIV). Small animals not included
(16, 17) Christodoulakis et al. (1996, p 212)

fall in agricultural production brought about by the forced displacement of villagers away from their cultivations, as discussed in Chap. 2.

3. Capital stock:
   With subscript PW denoting the prewar period 1934–1938, the proportion of destroyed factories is written:

$$\frac{dN_{CW}}{N_{CW}} = \frac{dN_{CW}}{dN_{PW}} \cdot \frac{N_{PW}}{N_{CW}} \cdot \frac{dN_{PW}}{N_{PW}} = \frac{dN_{CW}}{dN_{PW}} \cdot \frac{V_{PW}}{V_{CW}} \cdot \frac{dV_{PW}}{V_{PW}} \tag{8.14}$$

Industrial output in 1949 had reached 90 % of its prewar level, thus $N_{CW}/N_{PW} = V_{CW}/V_{PW} = 0.90$.[1] Table 8.1 displays the annual growth rate of industrial output and the number of new factories that were set up during 1934–1938, that is (dV_{PW}/V_{PW}) and dN_{PW} respectively. An official report by the Communist Party reported that 241 factories were destroyed during hostilities, i.e. $dN_{CW} = -241$.[2] Thus all terms in (8.14) are known, and the loss of industrial capital stock can be calculated. It is found to be equal to −15.76 % of its initial level.

Elasticities of livestock and capital stock are approximated by the relative shares of their income in total value added. The share of non-agricultural capital income is obtained as $\eta = 0.360$ in 1954, the nearest period for which disaggregated data are available, while a similar calculation for agricultural income gives $\varepsilon = 0.315$; for details see Christodoulakis et al. (1996, p. 212). Substituting into (8.13), the total drop in output by the end of conflict is found to be −11.84 % or $\psi = -2.84\ \%$ per year in average during the period 1946–1949.

8.5.3 Loss Evaluation

Applying the Collier and Hoeffler (2007) time frame of recovery to the Greek Civil War, cumulative output losses are evaluated at present value in the beginning of the Civil War by the following formula:

$$LOSS = \sum_{t=1946}^{1970} \frac{Y_0 - Y_t}{(1+\rho)^{t-1946}} \tag{8.15}$$

The output index is set at $Y_0 = 100$ and then evaluated in each period by the growth identity $Y_t = Y_{t-1}(1 + g_t)$, while (ρ) denotes the discount rate. The growth rate is set at $g_t - \psi = -2.84\ \%$ for the civil war period $t = 1946–1949$ and afterwards by the formula:

$$g_t = \psi + b(1-\delta)^{t-1950} \tag{8.16}$$

The rebound effect is denoted by $b = 1\ \%$ and δ is the rate at which it peters out. Letting a period of 21 years for recovery, a rate of $\delta = 16\ \%$ is calibrated to imply that by 1970 the consequences of the Civil War had petered out.

[1] As reported by Stathakis (2002, p. 66).

[2] Rizospastis (2011, p. 564).

Discounted at $\rho = 5\,\%$ annually, total losses in (8.15) amount to 129 % of annual GDP.[3] The loss estimate exceeds the upper side of the confidence interval [90÷110 %] of cumulative GDP losses due to a relatively costly civil war, as found by Collier and Hoeffler (2007) to express. This is because the drop in Greek GDP was in average stronger than the −2.20 % annual growth loss found by the authors to accrue in other conflict-stricken economies.

8.6 Bullets Versus Books

As predicted by the endogenous growth model in Sect. 8.2, the incidence of a civil war has two growth-impeding effects:

First, it causes fiscal strain on the economy, as public expenditure has to expand to accommodate increasing demand for protection and security services. Higher public deficits are either left unattended and lead to an accumulation of public debt or financed by a rise in distortionary taxation, which, in turn, suppresses long-term growth.

Second, it leads to a reallocation of public resources away from investing in productive infrastructure to spending on defense and protection services. This is strongly manifested in Greece during and after the Civil War. Figure 8.3 shows that until 1951 more than 50 % of the Government Budget was financing the military, and only a meager 5 % was left for education expenditure. While all other European nations were trying to boost education and upgrade the formation of human capital in the postwar period, Greece was having different priorities.

The situation started to change only by the late 1950s, as the share on defense declined and that on education increased somewhat, but it was still inadequate to meet the overwhelming demand for better education, which was considered to be the most accessible channel for social improvement. A strong student movement sparked in the early 1960s, demanding the increase of this share to 15 % of the total budget, a target that was finally achieved in 1964 after becoming a central issue in the election campaign. However, the euphoria was not to last. The defense share went on the rise again in 1967 after the imposition of a military dictatorship in Greece, while that on education fell behind.

---

[3]Averof-Tositsas (2010, p. 385) asserts that material destruction due to the Civil War amounted to USD 250 million at 1948 prices. By further adding damages in dwellings, refugees' costs and labor time forgone, the loss rises to USD one billion at 1970 prices. The 1970 exchange rate was 30 Drachmas/$ and US CPI 1970/1948 was 1.61. Thus the additional cost of the Civil War was USD 621 million at 1948 prices. Greek GDP in 1948 was Drs 63,706 million or USD 1319 million at 1948 prices. The estimate represents 66 % of annual GDP in 1948 or 89 % of GDP in 1946, substantially below the present estimate. The difference is due to the fact that Averof-Tositsas leaves out of the calculation the losses in human capital due to forced expatriation and imprisonment.

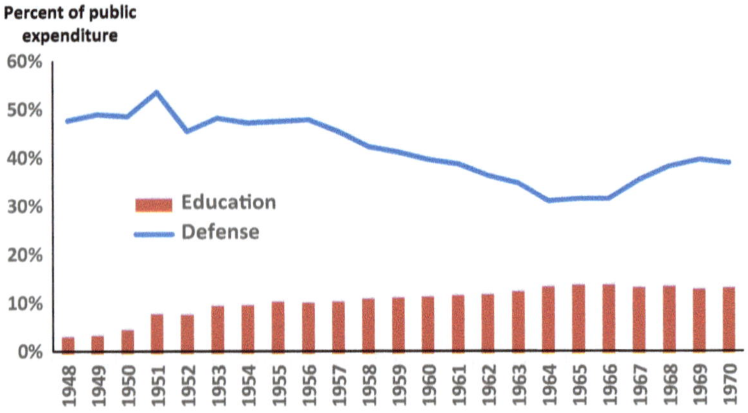

Fig. 8.3 Shares of public expenditure, Greece 1948–1970. *Notes:* A surge in defense spending occurred again in 1967, when a military dictatorship was imposed in Greece. *Source:* Kalafatis and Prontzas (2015). Table 85, p. 299

No wonder then that a massive wave of emigration, mainly to the US and West Germany, took place in the 1950s and 1960s, as hundreds of thousands were seeking a better opportunity abroad. The domestic labor force was reduced at a critical stage of development, when the country was in acute need of modernization in order to speed up assimilation with the nascent European Community.

Although the emigration effect was not taken into account in the cost calculation carried out in the previous Sect. 5, it is clear that such an outflow of active population has been—at least partly—caused by the Civil War and the authoritarian political system that prevailed afterwards. Though the examination of emigration effects on postwar economies is outside the scope of this book, it is perhaps worth noting that the flux of labor into northern European states, apart from Greece, was mainly coming from Spain, Italy, and Portugal. As a matter of fact, the first two had just experienced a civil war at that time, while the third was harshly repressed by a long-lasting dictatorship.

8.7 Counterfactual Analysis

8.7.1 The Cost of Conflict Escalation

In addition to the estimation of losses caused by the Civil War, the growth-accounting framework can also be used to assess the effect that counterfactual developments could have had on the conflict burden. As already described in Chap. 2, the conflict was terminated only after a major offensive against guerrillas took place in the summer of 1949. The scale of the operation was unprecedented and so were its consequences in terms of battle casualties, material resources and overall destruction.

The particular cost associated with the final offensive can be assessed in two steps as follows: first, evaluate the losses that would have accrued in the hypothetical case that an end of hostilities was negotiated well before the final offensive was launched. Then, by comparing the losses in the counterfactual and the actual case, an estimate of the escalation cost is obtained.

Assuming that an end to hostilities was negotiated in June 1949, the following alternative developments would have occurred instead of the actual ones:

(a) Casualties would be lower by those perished in the battles of July and August 1949, and similarly for the number of seriously wounded.
(b) Expatriation would have been altogether avoided.
(c) Imprisoned political rivals would be released and have re-entered active population soon afterwards. Given that most of the executions had already been carried out by June 1949, the assumption does not apply to those sentenced to death.
(d) The destruction of capital stock and livestock is assumed to be the same as in the actual case, since the final confrontation took place in the mountains and had little direct effect on urban or rural establishments. Therefore, the rate at which the post-conflict rebound peters out is also assumed to be the same.

As seen in the last column of Table 8.1, there would have been 117,139 fewer human losses, or 51 % lower than the number of losses that actually occurred. Active population would be enhanced by their participation, and economic activity could recover more quickly. Under these circumstances, the Civil War effect on GDP would peter out in the early 1960s, instead of 7 years later, as shown in Fig. 8.4 Following the same accounting framework as before, GDP losses are now evaluated at 90 % of annual GDP. Therefore, the cost produced by the final offensive in

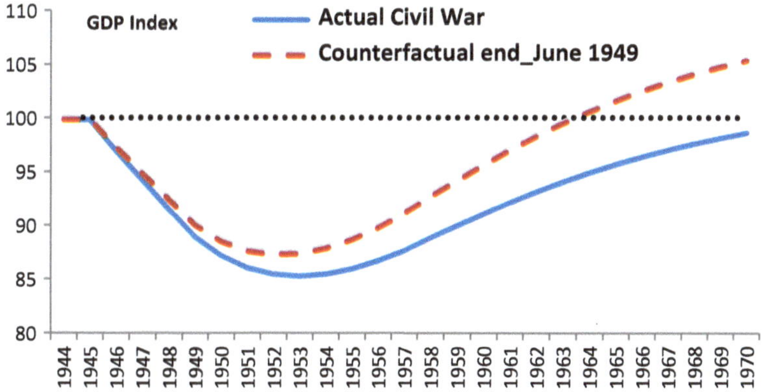

Fig. 8.4 A schematic representation of GDP recovery. (a) With Civil War casualties and persecutions as actually happened. (b) Counterfactual termination of hostilities in June 1949

summer 1949 is estimated at the tune of 39 % of GDP, nearly a third of total losses. Other counterfactual exercises may similarly be quantified.

Appendix

The Endogenous Growth Rate in an Economy with Civil War

The growth rate of output is endogenously derived for a closed economy along the lines described by Barro and Sala-i-Martin (1995, Ch. 5). To conform to the time framework presented in Chaps. 5 and 7, the model is specified in discrete time. For simplicity, population growth is assumed away. Hence, we consider an economy populated by a constant number (N) of identical households, each owning a firm with a capital stock (k) that depreciates at rate (δ). The government imposes a distortionary tax rate (τ) and uses collected revenues to finance productive infrastructure and enforce property rights, as described in Sect. 2.1.

Production per firm is characterized by constant returns to scale and given by the linear function:

$$y = A \cdot k \cdot \pi(\omega\tau) \cdot h[(1-\omega)\tau] \tag{8.17}$$

The net real rate of return to capital is (ρ_k), thus the rental cost of the firm's stock is $(\rho_k + \delta)$. Profit maximization requires that the after-tax marginal product of capital equals the rental cost, thus

$$(1-\tau)(\partial y/\partial k) = (1-\tau)A\pi h = \rho_k + \delta \Longrightarrow \rho_k = (1-\tau)A\pi h - \delta \tag{8.18}$$

Output, consumption, and capital stock per household grow at the same rate that is endogenously derived below. Households choose consumption in order to maximize a discounted utility function given by

$$\max_{c_t} U = \sum\nolimits_{t=1}^{\infty} \frac{1}{(1+\rho)^t} \cdot \frac{c_t^{1-\varphi}}{1-\varphi} \tag{8.19}$$

where ρ is the rate of discounting the future, and φ is a risk aversion parameter.

Non-consumed output is invested in the firm. The capital accumulation constraint is:

$$k_{t+1} = (1-\delta) \cdot k_t + (1-\tau) \cdot y_t - c_t \tag{8.20}$$

The transversality condition requires that

$$\lim_{t \to \infty} k_t (1+\rho)^{-t} = 0 \tag{8.21}$$

The Hamiltonian at period (*t*) is defined as:

$$\mathcal{H}_t = \frac{c_t^{-\varphi}}{1-\varphi} + \upsilon_{t+1} \cdot [(1-\delta)k_t + y_t - c_t] \tag{8.22}$$

where υ_{t+1} is the Lagrange multiplier for next period's capital stock. The first-order conditions for maximization are

$$\frac{\partial \mathcal{H}_t}{\partial c_t} = 0 \Rightarrow \upsilon_{t+1} = c_t^{-\varphi} \tag{8.23a}$$

and

$$(1+\rho) \cdot \upsilon_t = \frac{\partial \mathcal{H}_t}{\partial k_t} = \upsilon_{t+1} \cdot \left[1 - \delta + (1-\tau) \cdot \frac{\partial y_t}{\partial k_t}\right] \tag{8.23b}$$

The second order condition is satisfied as a consequence of decreasing marginal utility, i.e.

$$\frac{\partial^2 \mathcal{H}_t}{\partial c_t^2} = -\varphi c_t^{-\varphi-1} < 0 \tag{8.23c}$$

Lagging (8.23a) one period and substituting into (8.23b), we obtain:

$$\frac{\upsilon_t}{\upsilon_{t+1}} = \left[\frac{c_t}{c_{t-1}}\right]^{\varphi} \tag{8.24}$$

The growth rate of consumption per household—and similarly that of output and capital stock—is finally obtained as

$$g = \frac{c_t - c_{t-1}}{c_{t-1}} = \left[\frac{(1-\tau)A\pi h + 1 - \delta}{1+\rho}\right]^{1/\varphi} - 1 \tag{8.25}$$

For relatively small values of ρ and δ, the following approximation is obtained.

$$g \approx \frac{1}{\varphi} \cdot [(1-\tau)A\pi h - \delta - \rho] \tag{8.26}$$

This is the same as that obtained by Barro & Sala-i-Martin for continuous-time models.

Proof of Propositions 1, 2, 3

Taking the total differential of Eq. (8.8a) we get

$$(1+\beta)\cdot d\tau = -\frac{d\tau}{1+\lambda(1-\tau)} - \left\{\frac{ln[1+\lambda(1-\tau)]}{\lambda^2} - \frac{(1-\tau)}{\lambda\cdot[1+\lambda(1-\tau)]}\right\}\cdot d\lambda$$

or

$$\left(1+\beta+\frac{1}{1+\lambda(1-\tau)}\right)\cdot d\tau = -\frac{1}{\lambda^2}\left\{ln[1+\lambda(1-\tau)] - \frac{\lambda(1-\tau)}{1+\lambda(1-\tau)}\right\}\cdot d\lambda$$
$$= -\frac{1}{\lambda^2}\cdot \Psi(x)\cdot d\lambda \quad (8.27)$$

where for brevity it was set $x=\lambda(1-\tau)$ and

$$\Psi(x) = ln[1+x] - \frac{x}{1+x} \Rightarrow \Psi'(x) = \frac{x}{(1+x)^2} > 0, \text{for} x > 0 \quad (8.28)$$

Thus function $\Psi(x)$ is increasing and, given that $\Psi(0)=0$, it is positive for every $x>0$. Then from (8.27) it easily follows that

$$\frac{d\tau}{d\lambda} = -\frac{1}{\lambda^2}\cdot\left(1+\beta+\frac{1}{1+x}\right)\cdot^{-1}\Psi(x) < 0 \quad (8.29)$$

and Proposition 1 is established. From (8.8b) we obtain

$$\omega\tau = (1+\beta)\tau - 1 \quad (8.30)$$

Differentiating w.r.t. (τ), we get $\partial(\omega\tau)/\partial\tau > 0$ and $\partial\omega/\partial\tau > 0$, hence Propositions 2 and 3 are readily obtained.

References

Alesina A, Perotti R (1996) Income distribution political instability, and investment. Eur Econ Rev 40(6):1203–1228. doi:10.1016/0014-2921(95)00030-5

Averof-Tositsas E (2010) By fire and axe: Greece 1946–1949 and the precedents. Estia Editions, Athens (in Greek)

Barro R, Sala-i-Martin X (1995) Economic growth. McGraw-Hill, New York

Christodoulakis N, Dimeli S, Kollintzas T (1996) Economic fluctuations and growth in Greece and Europe. Stamoulis Editions, Athens (in Greek)

Collier P, Hoeffler A (2007) Civil war. In: Sandler T, Hartley K (eds) Handbook of defense economics, vol 2. Elsevier, Amsterdam, pp 712–737

Economides G, Park H, Philippopoulos A (2007) Optimal protection of property rights in a general equilibrium model of growth. Scand J Econ 109(1):153–175. doi:10.1111/j.1467-9442.2007.00486

Gupta S, de Mello L, Sharan R (2001) Corruption and military spending. Eur J Polit Econ 17(4): 749–777. doi:10.1016/S0176-2680(01)00054-4

Imai K, Weinstein J (2000) Measuring the economic impact of civil war. Centre for International Development, Harvard University, Working Paper No. 51, June

Kalafatis T, Prontzas E (2015) Economic history of the Greek state. In: Data of quantitative analysis, vol III. Cultural Foundation Publications, Piraeus Bank, Athens

Michiotis N (2007) In the name of the king: emergency martial courts in Greece 1946–1960. Synchroni Epochi Editions, Athens (in Greek)

Murdoch J, Sandler T (2002) Economic growth, civil wars, and spatial spillovers. J Confl Resolut 46(1):91–110. doi:10.1177/0022002702046001006

Murshed SM (2010) Explaining civil war: a rational choice approach. Edward Elgar, Cheltenham, UK

Papathanasiou I (2002) Arms by the side: verbal war or reconstruction policy? In: Nikolakopoulos E, Rigos A, Psallidas G. (eds.) The civil war: from varkiza to grammos (February 1945–August 1949), edited Proceedings, Themelio Editions (in Greek), Athens, pp. 143–161

Rizospastis (2011) The three-year epic of the Democratic Army of Greece 1946–1949. Synchroni Epohi Editions, Athens (in Greek)

Stathakis G (2002) The economy during the civil war. In: Nikolakopoulos E, Rigos A, Psallidas G (eds) The civil war: from varkiza to grammos (February 1945–August 1949), edited Proceedings. Athens, Themelio Editions (in Greek)

Chapter 9
The Comparative Economic Cost of Civil Wars

Another approach for measuring the overall economic impact of a civil war is by creating a hypothetical trajectory that the economy could have taken in the absence of conflict, and then juxtaposing it with actual developments to obtain the cost incurred. The method involves the creation of a "synthetic control" from a properly weighted average of other regions that are otherwise similar to the inflicted area. With regards to national civil wars, such an assumption is far from being realistic, as other countries have different institutions and may be subjected to idiosyncratic shocks. This chapter describes a modified framework, where a synthetic control is setup by stages, with each stage including a specific set of shocks.

9.1 The Comparative Approach

Another approach for measuring the economic impact of a civil war is by creating a hypothetical trajectory that the economy could have had taken in the absence of conflict. This trajectory is then juxtaposed with actual developments to obtain an estimate of the cost incurred. In this way, Abadie and Gardeazabal (2003, AG hereafter) assessed the cost incurred in the Basque region, by projecting a weighted average of other Spanish regions and comparing it with actual developments. The underlying assumption in their model is that all Spanish regions are institutionally similar and subject to symmetric shocks, thus they can be used as control variables to determine the conflict-free hypothetical path.

With regards to national civil wars, a group of other countries should be chosen as the control vector in order to obtain the hypothetical no-war trajectory. This may be daunting, as it is no longer realistic to assume that all members of the control group share the same institutions and are subject to the same war-unrelated shocks as the countries engaged in the conflict. In practice, countries may have idiosyncratic institutions and also are subject to asymmetric shocks. This calls for a number

N. Christodoulakis, *An Economic Analysis of Conflicts*,
DOI 10.1007/978-3-319-32261-2_9

of modifications in the AG approach, so as to obtain the comparative cost due to the Greek Civil War.

9.2 A Two-Stage Approach

To estimate the macroeconomic impact of the Greek Civil War in terms of output forgone, a hypothetical path of potential GDP (Y^{HP}) is obtained first and then compared with the actual course of the economy (Y^{GR}).

Similarly as with the AG model, the idea is to construct a "synthetic Greece" by using other European nations as controls. However, this is hampered by the fact that the major shock of WW2 has affected them in a highly asymmetric way: some of them were aggressors, other remained neutral, and even those that were occupied suffered at a different scale. Moreover, the communist system that prevailed after 1945 in Eastern Europe meant that these countries are unsuitable for describing a hypothetical trajectory for Greece, given that—after all—the country remained in the Western bloc.[1]

To account for the above, the AG framework is modified and a two-stage approach is adopted to obtain a synthetic control for Greece as follows:

9.2.1 *Estimation Stage I*

To delineate the effects of WW2 from the Greek Civil War, two vectors $\boldsymbol{Z}$ and $\boldsymbol{X}$ measuring GDP of Western European countries are considered. The first group ($\boldsymbol{Z}(j)$, $j = 1 \ldots 4$) consists of Sweden, Portugal, Ireland, and Switzerland, which remained neutral during WW2.[2]

The second group ($\boldsymbol{X}(i)$, $i = 1 \ldots 4$) includes Belgium, Denmark, Netherlands, and Norway, which—like Greece—were invaded by the Axis powers in 1940 and subsequently remained occupied until the first months of 1945.[3] France is treated as a special case because of its much larger size, the fact that WW2 involved its colonies, and the peculiarities of the Vichy regime.[4] Alternative estimates with and without France ($i = 5$), will be obtained for comparison.

[1]Eastern European countries would be chosen as "synthetic controls" for Greece in case the Civil War had been won by the Communist Party.

[2]Spain and Finland are excluded from both groups because they were allied with Germany for a substantial period of WW2. UK is left out of the second group because it was not occupied during WW2.

[3]Denmark, Norway, and the Netherlands were liberated in May 1945. Most of Belgium got free by the end of 1944 and the remainder in early 1945. Mainland Greece was liberated in October 1944, but Crete remained under German control until June 1945.

[4]The southern part of France was under special collaborationist arrangements with Germany until it was fully occupied in 1942. Furthermore, France in the 1950s was involved in two external conflicts, Indochina and Algeria.

The above selection is justified by the fact all nine countries of groups ***X*** and ***Z***—as well as Greece—shared some crucial economic characteristics both before and after WW2. These included:

(a) All of them joined the interwar Gold Exchange System and by the late 1930s had abandoned it;
(b) Postwar, all countries took part in the Bretton Woods arrangements and, eventually, all but Switzerland became full members of the European Union[5];
(c) All occupied countries $(i = 1,\ldots,5)$ received reconstruction aid through the US Marshall Plan, which helped recovery and catalysed economic cooperation.[6]

Both groups are supposed to have faced a common shock, stemming from the international environment throughout the period of examination. In the meanwhile the second group had suffered from invasion and occupation until 1945, and then experienced the post-liberation rebound of 1946–1947. Both effects are considered proportional to the aggregate human losses incurred during either invasion or occupation, and expressed as a ratio to prewar population.[7] Table 9.1 shows human losses to vary substantially across Nazi-occupied Western countries, thus the war shock, say vector ***w(i)*** with $i = 1\ldots5$, is clearly asymmetrical.

The Marshall Plan turned out to be another source of differentiation in the postwar period. Initially, the Plan was conceived to be distributed roughly proportionally to population,[8] but as shown in Table 9.1 there were substantial disparities between recipient countries. Hence, postwar aid is also treated as an asymmetric shock, say vector ***m(i)***.

Using the neutral group ***(Z)*** as a control for the second, the following cross-section equation is estimated for the GDP growth rates in occupied Western countries:

$$\Delta lnX^i = c^i + \beta_1 \Delta \ln\left[n^T \cdot Z\right] + \beta_2 DW \cdot w^i + \beta_3 DR \cdot w^i + \beta_4 m^i + \varepsilon^i \qquad (9.1)$$

In the above expression, Δln is the log-difference operator, superscript T denotes transposition, ε^i is an error term, and (c^i) is a country-specific effect ($i = 1,..,4$ or 5 when France is included).[9] GDP data are in 1990 US dollars from the Maddison historical series. The effect of WW2 is captured by a dummy $DW = 1$ for the period 1940–1945 and zero elsewhere. Similarly, dummy $DR = 1$ for the period 1946–1947 is used to capture the postwar rebound. The implementation of the

[5]Norway entered the then European Community in 1969 and exited in 1973.

[6]For a detailed account on the Marshall Plan, see Price (1955), Wexler (1983), Gardner (2001), and Bossuat (2008) among many others.

[7]From Fig. 1, it is apparent that the rebound in each economy was roughly symmetric to its suppression during the war.

[8]As noted by Price (1955).

[9]If scaled by GDP per capita at the beginning of the estimation period, these constants denote convergence effects towards the control group of neutral countries.

Table 9.1 War toll and the Marshall Plan

	War toll in Western Nazi-occupied countries[a]					
	BEL	DEN	NLD	NOR	FRA	Greece
Battle-deaths	12,100	2100 3900[b]	17,000	3000	217,600	35,100
Non-battle human losses	75,900	1100	284,000	6500	350,000	300,000
Total war toll as % of 1939 population	1.05 %	0.19 %	3.45 %	0.32 %	1.36 %	4.64 %
	Marshall Plan[c]					
Bossuat (2008), $ mio	546.6	257.4	877.2	236.7	2444.8	551.1
Price (1955), $ mio	556	271	1079	254	2706	694
Average, $ mio	*551.3*	*264.2*	*978.1*	*245.4*	*2575.4*	*622.6*
Population 1947 ('000)[d]	8450	4146	9630	3165	40,680	7529
Aid, $ per head	65.2	63.7	101.6	77.5	63.3	82.7

[a]Frumkin (1951)
[b]The toll for Denmark includes the losses of the pro-Nazi Danish Freikorps that fought in the Eastern Front and suffered 3900 battle deaths. Data from Danish Military History website milhist.dk
[c]Bossuat (2008, Fig. 1.1, p 14); Price (1955, Fig. 2, p 90)
[d]Maddison historical series

Marshall Plan is assumed to take place in three equal installments over 1948–1950, since more allocation details are lacking.

Control countries are weighted by vector ***n(j)***, j = 1...4. In Abadie and Gardeazabal (2003) weighting is chosen so as to reflect an approximation of some growth predictors in the synthetic region by analogous indices in the control group.[10] Here, however, no such set of indicators is available, thus in the absence of a better criterion weighting is set as simple averaging, i.e. $n(j) = 0.25$, $j = 1...4$. Regarding the coefficients to be estimated, a value of (β_1) close to one would suggest similar growth patterns across Western European countries.

As discussed in Chap. 6, to account for possible heteroskedasticity and contemporaneous correlation in the error terms, estimation is obtained by the SUR method, which is found to have the best explanatory power among alternatives. Depending on whether France is included or not in the control group, two alternative estimations are displayed in Table 9.2. All coefficients have the expected sign and are statistically significant at the 1 % or 5 % level. To see the quality of approximation by (9.1), the panel in Fig. 9.1 displays the dynamic forecast of GDP over 1931–1960 versus actual trajectories of the four countries.

The postwar path is relatively well represented, without a significant systematic deviation in any of the four controls. In the prewar period, a systematic deviation appears for the Netherlands, where the *ex post* GDP forecast is noticeably higher than actual. The explanation is that the Dutch economy took too long to exit the

[10]Apart from availability of data, the choice of controls is not straightforward in practice. Even with many compatible regions, Abadie and Gardeazabal (2003) finally choose only two of them as the control group.

Table 9.2 Estimation and forecasting of 'synthetic controls'

Dependent variable → Independent variable ↓	100*Δ[ln(*X*)]	
OLS, fixed effects (S.E. in brackets)	4-countries controls	5-countries controls (incl. France)
Constant	−0.4245 (0.600)	−0.388 (0.604)
$\beta_1 100^*\Delta[\ln(n^T Z)]$	1.098 (0.112)***	1.106 (0.109)***
β_2 *[occupation]*	−23.17 (4.890)***	−24.04 (5.30)***
β_3 *[postwar rebound]*	20.79 (7.793)***	22.12 (8.45)**
β_4 *[Marshall Plan]*	0.14 (0.039)**	0.135 (0.038)***
Nobs	$4 \times 25 = 100$	$5 \times 25 = 125$
R^2 adj	0.652	0.598
S.E. of regression	1.002	1.005
F-stat (prob)	27.51 *(p = 0)*	24.06 *(p = 0)*
DW	2.273	1.992
Greece-specific constant	1.167	1.339
S.E. 1931–1946	6.36	6.14
Wald test for $\beta_1 = 1$	$p = 0.38$	$p = 0.34$
Discounted loss 1946–1960 as % of 1946 GDP	124.87 %	166.80 %

Notes: Standard errors in brackets. One, two or three stars indicate p-values at 10 %, 5 % or 1 % respectively. The discount rate in the last row is set at 5 % discount p.a. Time dummies are set to 1 at the following periods: for war and occupation 1940–1945, for the rebound 1946–1947, and for the Marshall Plan 1948–1950. *Source:* Data as defined in Appendix

Gold Standard and then was trapped in a prolonged stagnation that ended just before WW2; for a discussion on this issue, see de Jong and van Zandem (2014).

9.2.2 *Estimation Stage II*

The synthetic group of Nazi-occupied countries is now used as a control to obtain a "synthetic Greece" and the hypothetical GDP trajectory in the absence of civil war. Potential GDP for Greece (Y^{HP}) is calculated in each period 1946–1960 by substituting for the Greek war toll and per capita aid to Greece into estimated equation (9.1).

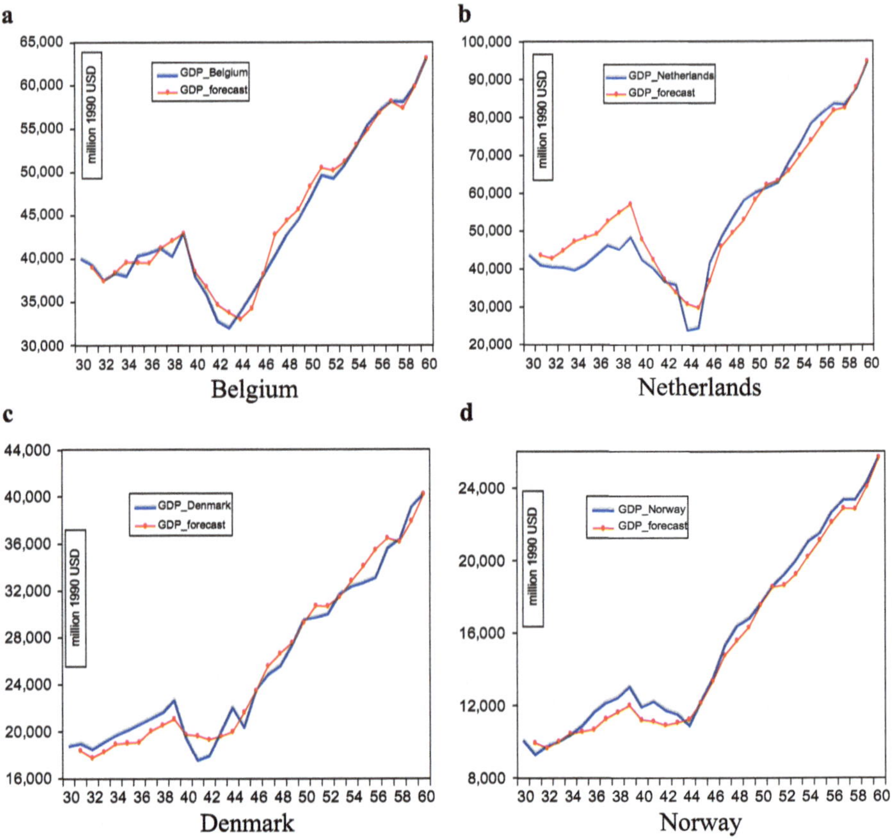

Fig. 9.1 Actual and forecasted GDP for Nazi-occupied countries (**a**) Belgium, (**b**) Netherlands, (**c**) Denmark, (**d**) Norway

Using a hat to denote estimated coefficients, potential GDP is obtained as:

$$\Delta lnY^{HP} = c^{GR} + \widehat{\beta}_1 \Delta \ln\left[n^T \cdot Z\right] + \widehat{\beta}_2 DW \cdot w^{GR} + \widehat{\beta}_3 DR \cdot w^{GR} + \widehat{\beta}_4 m^{GR} \qquad (9.2)$$

The country-specific effect (c^{GR}) is chosen for Greece so as to minimize the squared residuals for the period prior to the outbreak of the Civil War. Specifically:

$$c^{GR} = \text{argmin} \sum\nolimits_{t=1931}^{t=1946} \left[Y_t^{HP} - Y_t^{GR}\right]^2 \qquad (9.3)$$

The difference ($Y^{HP} - Y^{GR}$) is finally taken as a proxy for the cost of the Civil War, and is shown by the shaded area in Fig. 9.2. While other European nations were experiencing a fast recovery after 1946, Greece was losing momentum, and growth remained low until 1952. It was only after a drastic devaluation of the Drachma by

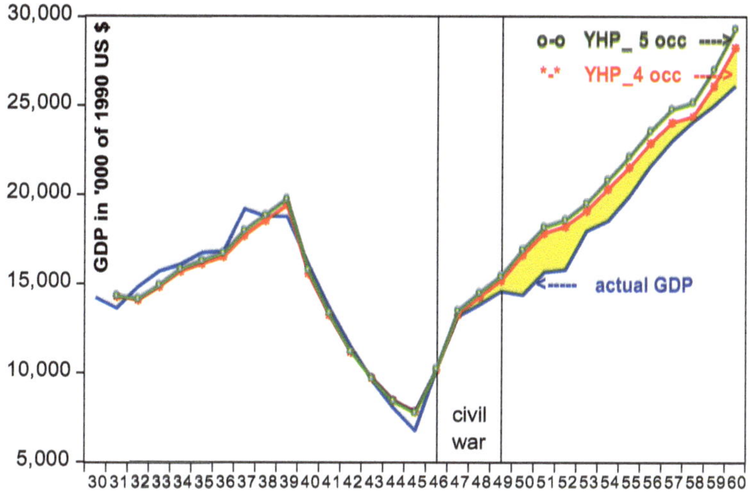

Fig. 9.2 Actual and potential GDP for Greece 1946–1960. Actual GDP is denoted by simple line; the hypothetical GDP estimated by using four occupied countries (YHP_4) is denoted by asterisks; the one estimated by using five occupied countries (YHP_5) by circles. GDP is expressed in thousands of US Dollars at 1990 prices. The shaded area is a measure of the failure of Greece to follow other European states in postwar reconstruction

50 % in 1953 and the subsequent reduction in real wages that growth accelerated and activity reached normal levels by the end of the decade.

The two trajectories seem to converge in 1958, but then a new surge appears in the synthetic path, due to the fact that Belgium and the Netherlands stepped up growth after co-founding the European Community in 1958 with France, Italy, Luxembourg, and Germany. Greece was not yet ready to participate and had to wait for another 3 years to apply for a Connection Agreement. But even that was put on hold after the imposition of dictatorship in 1967 and Greece formally joined the European Community in 1980.

9.2.3 *Loss Evaluation*

Following Collier and Hoeffler (2007, p. 40), economic activity is assumed to take about 10 years to recover after a civil war, so the cumulative loss is calculated in present value until 1960 by the formula:

$$LOSS = \sum_{t=1946}^{t=1960} \frac{YHP_t - YGR_t}{(1+r)^{t-1946}} \tag{9.4}$$

The discount rate is set at 5 % per annum. When France is not included in the control group, cumulative output losses are found to be 125 % of 1946 GDP, which is taken as the starting year of the Civil War. With France included, losses rise to 167 % of GDP. The reason for the latter to be higher is the fact that estimation now takes into account the particularly strong growth witnessed in postwar France.[11]

The simple average of the two estimates is 146 %. This is substantially above the loss of 129 % of GDP based only on the destruction of production factors. This 17 % difference may be attributed to the omission of the destruction of infrastructure in the evaluation of Sect. 8.5, for which no reliable data can be found. For example, Babanassis (2001) reports that 15 % of railway lines were destroyed during the conflict, but there is no calculation available.

Whatever the most accurate estimate, the cost of the Greek Civil War was terrifyingly high.[12] Next question is why such a costly confrontation went out of control for so many years, and this is briefly discussed in the next chapter.

References

Abadie A, Gardeazabal J (2003) The economic costs of conflict: a case study of the Basque country. Am Econ Rev 93(1):113–132

Babanassis S (2001) The economic aspects of the civil war. In: Koutsoukis K, Sakkas I (eds) Aspects of civil war 1946–1949. Filistor Editions, Athens, pp 39–54 (in Greek)

Bossuat G (2008) The Marshall Plan: history and legacy. In: Sorel E, Padoan PC (eds) The Marshall Plan: lessons learned for the 21st century. OECD, Paris, September

Collier P, Hoeffler A (2007) Civil war. In: Sandler T, Hartley K (eds) Handbook of defense economics, vol 2. Elsevier, Amsterdam, pp 712–737

de Jong H, van Zandem JL (2014) Debates on industrialization and economic growth in the Netherlands. Department of Economic History, Lund University, Seminar Paper no. 113

Frumkin G (1951) Population changes in Europe since 1939. Geneva. http://en.wikipedia.org/wiki/World_War_II_casualties

Gardner R (2001) The Marshall Plan fifty years later: three what-ifs and a when. In: Schain M (ed) The Marshall Plan: fifty years after. Palgrave, New York, pp 120–129

Price HB (1955) The Marshall Plan and its meaning. In: Governmental affairs institute. Cornell University Press, Washington

Vetsopoulos A (2007) The economic dimensions of the Marshall Plan in Greece 1947–1952. Gutenberg Editions, Athens (in Greek)

Wexler I (1983) The Marshall Plan revisited: the European recovery program in economic perspective. Greenwood Press, Westport, CT

[11]For this reason, the 30 years from 1945 to 1975 are popularly called *Les Trente Glorieuses*.

[12]Vetsopoulos (2007, Table 1) reports that, during 1947–1952, the Economic Cooperation Administration (ECA) disbursed to Greece USD 949 million. This amount corresponds to USD 126 per head and would have led to a synthetic GDP path much higher than the one previously considered. One should note, however, that the extra aid to Greece was earmarked to face the consequences of the Civil War, and most likely would not have been disbursed in its absence.

Chapter 10
A Critical Appraisal

The final chapter discusses how the use of dynamic models helps in understanding the escalation and duration of a conflict, as opposed to cross-section estimates based on sparse data from different countries. This facilitates the analysis of factors that may stand responsible for the break-out and the escalation of a conflict. Usually, these factors range from the lack of strategy and leadership to incomplete information, and the economic analysis can be used to determine the costs incurred by failure to reach a compromise early on. Greece is found to have been singularly prone to most of such factors during the 1940s and this explains the dire consequences that the Civil War had on the country.

10.1 The Role of Economic Analysis: A Postscript

The use of economic analysis to model and describe the process of a civil war is helpful in explaining various aspects and effects of the conflict. In the preceding chapters three main areas were investigated:

First, how a conflict evolves over time in terms of battle-casualties and army formations. Of particular importance was the existence of repetitive patterns in battle-casualties without a definite conclusion.

Second, the investigation of socio-economic conditions as well as other political and terrain factors as determinants of political discontent related to the onset and continuation of civil war.

Third, the calculation of the economic costs imposed by a civil war on the country's output and the long-term rate of growth.

The analysis included the setting-up of theoretical models that were subsequently empirically tested and estimated by using appropriate time-series. Extensive applications on the Greek Civil War were made possible by the availability of high-frequency data regarding battle-outcomes and socio-economic indicators. The possibilities and limitations of the above methods are summarized below.

N. Christodoulakis, *An Economic Analysis of Conflicts*,
DOI 10.1007/978-3-319-32261-2_10

10.1.1 Dynamic Conflict Modelling

A civil war is by definition a process calculated to be divisive in its proclamation, evasive in its tactics, and destructive in its effect. Setting up a modelling framework to capture specific patterns of civil wars, seems to be a daunting task and is no wonder that so many methodologies are suggested by alternative schools of thought.

The static opportunity-cost models of conflict are based on the theory of contests, in which two opponents compete for limited resources. Each side chooses the size of the army to fight against the other, in such a way as to maximize the expected payoff. The conflict success function depends on a number of economic, political, geopolitical and geographic factors, which are subsequently tested against various characteristics of the conflict by pooling together different countries and wars.

The static contest framework is helpful in explaining how normative considerations of opportunities and costs determine the decision to fight, but offer no insight on how the conflict evolves over time after such a decision is taken. On the other hand, ad hoc dynamic models that depict interactions between competing populations are useful to determine the patterns of conflict over time, but lack a normative framework on why such a confrontation is considered in the first place.

In the previous chapters an attempt was made to combine both normative aspects and dynamic analysis within a single framework. First, by devising the Lotka-Volterra interactions as reaction functions in a Nash-Cournot framework, in which each side sets specific objectives. Dynamics reveal important insights regarding the patterns of conflict, the prospects of termination and the risks of being trapped in repetitive warfare to the detriment of both adversaries. This type of analysis helps determine the key developments of each conflict and explain the mechanism through which they are influenced by specific factors.

A more promising framework of analysis is to examine the dynamics of conflict in an intertemporal framework of battle-planning under the constraints of army logistics. Apart from making the model a lot more realistic in capturing real-life characteristics, the intertemporal optimization framework allows the examination of the long-term effects that a conflict is likely to bring about.

10.1.2 The Correlates of War

One of the most debated issues in the academic literature on modern civil wars is whether and to which extent pooling different countries and civil war episodes that took place in different periods of time leads to a valid assessment of how conflicts are related to certain socio-economic and political factors. For example, Newman (2014, p. 15) criticizes this "positivist" worldview by arguing that "societies, and thus conflicts, are unique" and warns on the derivation of conclusions based upon "the configuration of variables across many cases—which reflect different circumstances and contexts."

A more cautious way to analyze the importance of socio-economic and political factors on conflicts is to use regional data from the same country, and examine how political discontent or the mobilization of insurgents is affected. The regional estimates are subject to the same institutional, political and geopolitical influences prevailing in the country as a whole, and, therefore the criticism of unifying different contexts does not apply.

Moreover, regional analysis in a single country allows for the differentiation between current and past grievances and how they influence the culture of discontent to develop over time. This analysis may be of critical importance on how reconciliation and reconstruction policies should be designed after the conflict is terminated so as to face the root causes of socio-political grievances.

10.1.3 Measuring the Cost of Conflict

By determining the severity of fighting in each phase of the conflict and the reasons that lead to the specific patterns of interactions between the adversaries, the dynamic analysis is also useful in assessing the cost incurred by the economy as a result of the failure to reach a compromise. The assessment takes into account how specific production factors are hit by the conflict and, thus, provides a more accurate evaluation than cross-country generalizations.

Subject to data availability, the economic loss imposed by the conflict can be evaluated at different periods of its duration and the results can be used to obtain an estimate of how much cost could have been avoided by an earlier termination of hostilities. Such estimates can be of particular interest when a counter-factual analysis is attempted to examine whether and how the conflict could have taken a different course in history or why conflict-generating factors prevail.

Economic objectives have been a central issue in too many conflicts worldwide, no matter if this is openly declared as the key motivation for participating in organized fighting or is hidden—by design or implicitly—underneath ideological, ethnic, or various religious agendas. The better these objectives are understood, the more are the chances that a compromise can be achieved between the adversaries before the conflict erupts or escalates beyond control.

The economic analysis of conflicts may help to delineate the relationships between socio-economic grievances and political discontent, and shed light on the factors that prevent a peaceful resolution of differences. The current approaches are still far from being uncontroversial or universally applied, but even so the analysis of civil wars, the design of prevention policies, and the plans of reconstruction pay increasingly more attention to their findings.

10.2 Conflict as a Country Failure

One of the most critical factors in shaping the history of a nation is how it responds to a devastating crisis that deprives a large number of citizens of the core of essential goods and services, and drives society to the brink of falling apart. A striking example of two opposite patterns observed in the twentieth century was Germany after the First World War and Germany after the Second World War. In the first case, a defeated and ruined nation followed a deeply divisive course, in which the elimination of enemies—both domestic and foreign—was set as a precondition for those prevailing to enjoy an unfettered prosperity. In the second case, a defeated and ruined nation endorsed a path of reconciliation towards both its own citizens and other nations. Though the country had to wait for nearly half a century for its own unification, the embrace-all policy made postwar Germany a model state with regards to social protection and economic prosperity. Besides, the model would survive many economic crises that were lying ahead.

In the aftermath of WW2, not all European nations had similarly conciliatory aspirations. In the first place, some countries did not even have the option, as they were either locked in prewar dictatorships—such as Spain and Portugal—or quickly subdued to authoritarian regimes—such as in Eastern Europe. In democratic countries in Western Europe, the way of settling entrenched rivalries between various social groups was far from simple or uniform. Most countries immediately realized the importance of political participation and social cohesion, and devoted plenty of resources to set up and finance social-inclusive systems, from welfare state to housing and education, in line with the conclusions reached by the analysis in Sects. 4.3 and 4.6. Countries under more confrontational circumstances, such as Belgium in 1944 and Italy in 1946, introduced constitutional guarantees for power sharing, in order to diffuse the threat of a civil conflict that was simmering since before and during the war.

In contrast, Greece appears to be a singular failure in the postwar map of Western Europe. Though it excessively suffered during the Occupation and despite its critical contribution to the Allied victory, the country was soon engulfed in yet another confrontation that crippled recovery and squandered the long-awaited peace dividend. The human losses in the Greek Civil War in fact exceeded the number of battle-deaths in the Italian and German invasions in 1940–1941, and imposed a huge cost on the economy, to the tune of more than an annual GDP. As a result, Greece missed the opportunity for a quick post-war recovery along the other European nations, and it took the country nearly a decade to return to normal economic activity.

Ex post, it seems obvious that an early termination of hostilities could have been proven beneficial for both sides, especially if one takes into account that the country had just exited another catastrophic war and prospects were naturally being expected to improve and provide more opportunities for all. But this didn't happen and is challenging to examine the reasons of such a failure.

10.3 Greek Failures

The reasons that explain the failure of the warring sides to agree on a peaceful resolution are several, and range from partisan opportunism and geopolitical miscalculations to the uncontrollable pressure by rank and file violence. The fact, however, that made such effects to be so critical is nothing else but the lack of sustainable strategy and strong leadership. The following aspects are examined:

10.3.1 Uncontrollable Violence

Uncontrollable violence was propagated by the rank and file, as both armies were recruiting rural outlaws previously involved in widespread looting. After WW2 several local militias joined either GNA or DAG in order to gain legitimacy and enjoy more seizure opportunities. DAG for its part had issued many warnings against looting the rural populations, while the Government had to send troops against local rightwing barons; for an account, see Gregoriadis (2011, p. 127). But as the Civil War was spreading, alertness by each side on its own banditry was soon subordinated to the aggressiveness being mounted against the other.

10.3.2 The Broken Trust

The Varkiza Treaty was seen by the Left as an act of humiliation after its supporters were first surrendering their arms and subsequently persecuted, and by the Right as mission unaccomplished with regards to annihilating its political rivals. Sfikas (2002) considers that during the Civil War KKE kept putting some extra conditions on its peace proposals as hedging against being cheated for a second time, and invariably this was giving the Government the pretext to reject it as unacceptable.

Such a characteristic episode took place when KKE made an appeal for ceasefire in May 1948 by publicly denouncing the seizure of power by force. The Government expressed some interest, but demanded that all children transferred by DAG to the People's Republics in Eastern Europe should return forthwith to Greece as "an unfailing test for the sincerity of the proposals."[1] The ultimatum was obviously rejected and a major offensive on mountain Grammos followed in a few weeks' time.

[1]For details see Averof-Tositsas (2010, p. 292).

10.3.3 Signaling by Proxy

In order to overcome the signaling breakdown, both sides attempted to lend credibility to their proposals by discussing it with the foreign patrons of the other side. Thus the Government approached the Soviet Union, asking it to intervene so as KKE abandons the armed struggle, while the Left developed regular contacts with the British; see Sfikas (2002, p. 85). But as the geopolitical landscape was entering the Cold War phase, such innuendos were quickly discarded for the following reasons:

First, the USSR was unwilling to get involved in arbitrating the Greek conflict and was instead trying to get completely disentangled and concentrate on its own security concerns. Second, the attitudes of the US and the UK had also been reversed. During the battle of Athens in 1945, the British were adamant about crushing the Left, while the Americans were openly supporting its participation to the Government as the only guarantee against the dominance of the Right. In the beginning of 1947, the US was favoring reconciliation as a precondition for the economic reconstruction and suggested that a "real amnesty" should be granted (Porter 2006, p. 229). Even after the Truman Doctrine was proclaimed, a sizable majority of Americans felt that Truman was wrong in bypassing the UN (Mazower 2012, p. 227).

As soon as the US replaced the British in the role of Greece's mighty guardian, both powers switched views on how the impasse should be resolved. In 1948, the British Government produced a plan for ending hostilities that incorporated several of the preconditions set by the communists, and it was now the turn of Americans to reject the mediations and opt for a clear victory over DAG. Besides, the British were sidelined in the geopolitics of the Cold War and ceased to be crucial players anymore, thus their delayed eagerness to reconcile Greece was not heard anymore by those in the frontline.

10.4 Missing Strategies

The reasons for failing to reach a consensus on power sharing and reconstruction can be traced to the mutual distrust that had prevailed for many years between the Left and the Right, their lack of a coherent strategy for state-building, and the inability to effectively negotiate an end to the conflict before it became too devastating. In a serial demonstration of political blunders, the leadership of both sides seemed to be solely preoccupied with its own dominance and survival, disregarding the long-term unity, cohesion, and wellbeing of the country. The main political deficiencies were the following:

10.4.1 The Communist Party: A Cause Without a Strategy

To start with, KKE was lacking a clear strategy on how to fight the conflict. The first question was on whether the Party was consciously engaged in the civil war or just reacting to adverse developments. Even several years afterwards, top party cadres continued to claim that a decision on war was indeed taken, while others insisted that this never happened. Confusion was openly manifested when KKE abstained from parliamentary elections in March 1946 to avoid legitimizing an allegedly rigged poll, but a few months later it took part in the restoration referendum that was indeed conducted with open violence against republicans.

Second, the Communist Party throughout the Civil War remained indecisive on whether to pursue a full urban confrontation, as advocated by its leader, or avoid it and instead engage in partisan action as favoured by the DAG chief. This soon led to adventurist tactics and—consequently—to pointless bloodshed for many years. From a military point of view, such opportunism was unforgivable as it ignored fundamental tactics. For example, Clausewitz (1976, p. 579) had long ago advised that "no one starts a war—or rather, *no one in his senses should do so*—without first being clear in his mind what he intends to achieve by that war and how he intends to conduct it" (emphasis added).

Eventually, the gap between the two views led to the dismissal of the guerrilla chief in January 1949. But, rather than making the leadership more realistic, this move widened the illusions about winning the conflict.[2] Whether driven by ignorance or otherwise, the pretension of superiority prevented conciliation efforts to gain momentum and thus contributed to the prolongation of conflict.

10.4.2 The Government: Power Without a Vision

The Government side was also experiencing a lack of robust leadership. Factionalism was the rule of the day and ten postwar Governments had been formed by the end of 1947, (i.e. with an average duration of less than 3 months each), thus making the search for a comprehensive strategy to look vain.

The unwillingness of the Government to deal with the mounting problems of reconstruction was omnipresent and the ministers' sole concern seemed to be how to be personally kept in office.[3] A chaotic situation would make them more likely to stay, because uncertainty usually blurs personal accounting and job demands. This allowed most of them to remain unaccounted for neglecting reconstruction and thus

---

[2]Perhaps the most notorious rejection of reality was the claim by the KKE leader 2 months *after* the final defeat in Grammos that "the main forces of DAG remain unscathed with arms by the side"!

[3]For a description of this alarming situation by the US mission to Greece, see the Report by Porter (2006, p 80 and 162).

perpetuating the grievances of the population. Moreover, small parties did not have a real interest in reconciliation, since the participation of the Left in the Government would have meant the termination of their own.

The hope that the political system would become more coherent under the aegis of the King withered quickly. Having spent most of his adult years either supporting a pro-fascist regime or staying abroad, the King showed no interest in inspiring national reconciliation or simply improving the life of Greek citizens; see Porter (2006, p. 83). His low-key successor seemed to be more consensual, but was quickly overtaken by his hardliner spouse and any notion of compromise evaporated quickly.

10.4.3 Lack of Leadership

Given the enormous human and economic cost of the war for both sides, it is clear that a negotiated end could have been to their mutual advantage, and the failure to do so must be explained not just by aggressiveness or cynicism, but mainly by the adversaries' inability to comprehend the situation and adopt a coherent strategy for their long-term objectives. A likely explanation for the severity and prolongation of the conflict is given by Brown's (1996) conclusion that bad leaders typically trigger violence not to resolve the existing problems or to preserve peace for their citizens, but to remain in power.

The lack of long-term strategies of reconciliation and co-existence both among the guerrillas and the government allowed hardliners to dominate both sides of the conflict, causing large and devastating effects on the country and its future. The conflict was followed by massive expatriation of defeated guerrillas and their families, while persecutions and political segregation lasted for another quarter century, before a full-fledged democracy was restored in 1974.

References

Averof-Tositsas E (2010) By fire and axe: Greece 1946–1949 and the precedents. Estia Editions, Athens (in Greek)

Brown ME (1996) The causes and regional dimensions of internal conflict. In: Brown M (ed) The international dimensions of internal conflict Cambridge. MIT Press, Cambridge, MA

Gregoriadis S (2011) N. Zachariadis: the fatal leader, vol IV. Kyriakatiki Editions, Athens (in Greek)

Mazower M (2012) Governing the world: the history of an idea. Allen Lane Editions, London

Newman E (2014) Understanding civil wars: continuity and change in intrastate conflicts. Routledge, London

Porter P (2006) Wanted: a miracle for Greece: diaries of a presidential envoy. Metamesonykties Editions, Athens (in Greek)

Sfikas T (2002) The ‘war-peaceful’ dimension in the Greek civil war: peaceful initiatives and compromise possibilities 1945–1949. In: Nikolakopoulos E, Rigos A, Psallidas G (eds) The civil war: from varkiza to grammos (February 1945–August 1949). Athens, Themelio Editions (in Greek)

von Clausewitz C (1976) On war. Howard M, Paret P (ed & transl). Princeton University Press, Princeton, NJ

Chapter 11
Appendix: Data Sources and Definitions

11.1 Nomenclature

Nomenclature of the Greek Civil War was never agreeable and each side was offensively labeling its opponents. For the Government, the army was the Greek National Army (GNA), while its rivals were "bandits", "robber-bandits" or even "Slav-gangs." The communists had proclaimed the Democratic Army of Greece (DAG) and brandished the Government as "imperialist lackeys" and its forces as "monarchist-fascist troops", as well as "robber-bandits" by reciprocation. The conflict itself was accordingly called "counter-bandit struggle" or "liberation struggle" by the Government and the communists respectively. It was mutually described as a "civil war" only in the 1980s. The present paper adopts a terminology as close as possible to each side's preferences for its own troops. Thus GNA stands for Government troops, army soldiers and state forces, while guerrillas, fighters, and rebel forces ("*andartes*") are interchanged in describing DAG.

11.2 Data Sources

ESYE, 1939. *Annuaire Statistique de la Grèce 1939*. Athens.

ESYE, 1946. *Population de la Grèce 1940*. Athens.

ESYE, 1955. *Population de la Grèce 1951*. Athens.

ESYE, 1958. *Results of Agricultural Census of Year 1950*. Athens.

GES, 1970, *The Greek Army in the Anti-guerrilla Struggle 1946–1949: The cleansing of Roumeli and the first battle of Grammos*. Athens, (in Greek).

GES, 1971, *The Greek Army in the Anti-guerrilla Struggle 1946–1949: The first year of the anti-guerrilla struggle 1946*. Athens, (in Greek).

GES, 1976, *The Greek Army in the Anti-guerrilla Struggle 1946–1949: Operations of the Third Army Corps 1947–1949*. Athens, (in Greek).

N. Christodoulakis, *An Economic Analysis of Conflicts*,
DOI 10.1007/978-3-319-32261-2_11

Table 11.1 Conflict data 1946–1949, in persons

	GNA	SKLD	SWND	SMIA	DAG	RKLD	RWND	RCAP	RSUR	ABDUCT	DISPL
Jan 1946		5	4	0		3	4	36			
Feb 1946		0	0	0		0	0	0			
Mar 1946		11	5	0		0	0	1			
Apr 1946		4	8	0		4	12	15			
May 1946		8	16	0		1	0	0			
Jun 1946		22	12	17		10	5	62	6		
Jul 1946		24	35	40		65	32	146			
Aug 1946		50	58	52		52	0	45	31		
Sep 1946		126	81	82		85	43	36	134		
Oct 1946		22	42	34		139	70	61	43		
Nov 1946		108	83	70		123	62	26	27		
Dec 1946		120	81	141		115	57	52	118		
Jan 1947	92,000	49	58	101	10,850	130	86	21	19	812	
Feb 1947		36	49	109	14,850	128	84	37	15	1508	
Mar 1947		65	74	72	16,250	280	143	31	7	794	28,651
Apr 1947		94	171	26	17,050	838	185	756	246	1062	
May 1947		92	127	115	16,450	804	381	477	249	852	
Jun 1947		61	170	21	16,900	608	271	517	362	1403	140,880
Jul 1947		151	324	110	16,900	1139	726	306	217	2085	
Aug 1947		92	206	155	16,700	357	167	63	28	1933	159,191
Sep 1947		46	85	22	17,400	269	76	72	30	2213	
Oct 1947		93	304	39	18,600	499	131	484	327	3047	
Nov 1947		121	434	44	18,600	541	320	210	115	2851	
Dec 1947		167	460	146	20,350	494	203	88	143	1491	
Jan 1948	120,098	176	383	264	22,250	835	2505	609	446	3650	270,727
Feb 1948	132,000	129	337	135	24,140	731	2193	553	486	1500	287,239

Mar 1948		266	685	230	25,700	1459	4377	968	999	2000	
Apr 1948		241	799	316	23,900	1371	4113	1059	599	1600	
May 1948		248	609	162	25,610	1301	3903	1527	931	810	
Jun 1948		321	983	345	23,300	1304	3912	579	579	1062	348,772
Jul 1948		601	2433	172	22,090	1817	5451	690	839	1404	346,831
Aug 1948		626	3258	158	21,100	1896	5688	634	701	3010	
Sep 1948		412	1854	207	23,720	1831	5493	635	833	2473	
Oct 1948		324	1241	167	25,480	1012	3036	332	619	3600	320,168
Nov 1948		294	875	455	25,450	891	2673	588	554	1924	
Dec 1948		205	618	192	24,000	1279	3837	741	670	1848	294,906
Jan 1949		225	681	271	24,090	1375	4125	657	931	1500	
Feb 1949		253	1222	792	21,810	1846	5538	1264	791	1730	
Mar 1949		140	426	37	19,450	1894	5682	2219	1263	319	
Apr 1949		275	1074	336	20,200	1924	5772	1034	1020	415	
May 1949		221	817	146	18,320	1161	3483	1133	1049	720	
Jun 1949		266	662	246	17,365	927	2781	1199	781	100	
Jul 1949		192	599	42	17,400	653	1959	696	573	143	
Aug 1949		771	3712	73	3580	2588	7764	1543	566	32	
Sep 1949					2410	551	1653	442	362	20	
Oct 1949					1910	173	519	450	464	11	
Nov 1949					1275	100	300	194	283	16	
Dec 1949	128,701				96	96	288	126	150	0	

Notes: For definitions of variables and data sources see Appendix

Table 11.2 DAG regional formations and allocation of persecutions

	Formation	Nearby mountain formations	Allocation of purges 1945–1946	Allocation of prosecutions 1946–1949	Purges 1945–1946	Prosecutions 1946–1948
1	Evia		23 % of FFE	Thiva	644	671
2	Parnassos	Vardousia Panetolikon	50 % of FFE	Athens, Lamia/2, Mesologi/2	1399	1671
3	Othrys	Magnesia	27 % of FFE	Lamia/2, Volos	755	1137
4	Agrafa	Tzoumerka, Xinias, Souli	Arta, Trikala, Karditsa	Trikala	1109	1136
5	Pindos	Smolikas, Orliakas Zagoria, Mourgana	Ioannina/2	Ioannina/2	4065	587
6	Hasia	Antihasia, Koziakas, Kamvounia	50 % Pieria & Larissa	Larissa/2	2000	1410
7	Olympos	Pieria, Ossa	50 % Pieria & Larissa	Larissa/2	2000	1410
8	Vermion	Siniatsiko, Vourinos	50 % of KGIP	Kozani/2	1417	1077
9	Grammos		Ioannina/2 Kastoria/2	Ioannina/2 Kastoria/2	5223	1636
10	Vitsi		Kastoria/2, Florina	Kastoria/2 Florina	3860	2625
11	Kaimktsalan	Paikon	50 % of KGIP	Kozani/2, Veria	1416	1077
12	Belles	Korona, Krousia		Kilkis		405
13	Halkidiki	Kerdylia, Pangaion		Thessaloniki/2		1651
14	Serres	Orvilos		Serres, Thessaloniki/2		1651
15	Haidu	Boz-Dag		Drama		1704
16	Thrace	Vyrsini, Sapka		Xanthi, Alexandroupoli		1744
17	Lesvos			Mytilene/2		94
18	Samos	Icaria		Mytilene/2		94
19	Hania	Lefka Ori		Hania		448
20	Cephalonia	Enos		Mesologi/2		0
21	Peloponnese			Patras, Tripoli, Corinth, Calamata		3083

(continued)

Table 11.2 (continued)

Formation	Nearby mountain formations	Allocation of purges 1945–1946	Allocation of prosecutions 1946–1949	Purges 1945–1946	Prosecutions 1946–1948
Total				23,888	25,308

Notes: FFE denotes the areas of Fthiotis, Fokis and Evia combined. Shares in local purges are set equal to the regional population shares in total population of Sterea (excluding Attica), according to the 1951 Census. KGIP denotes the areas defined by Kozani, Grevena, Imathia and Pella combined

GES, 1980, *The Greek Army in the Anti-guerrilla Struggle 1946–1949: The second year of the anti-guerrilla struggle 1947.* Athens, (in Greek).

JUSMAPG, *History, 1948–1950*, National Archives, US.

Michiotis N., 2007, *In the Name of the King: Emergency martial courts in Greece 1946–1960.* Athens: Synchroni Epochi editions (in Greek).

Grammenos, B. and P. Rodakis, (1987) That's how Civil War started. Reproduction of *"The 1947 DAG Report to the UN."* Athens: Glaros editions.

11.3 Battle Data

State army (GNA) figures include the Army, Gendarmerie, and armed nationalist groups. Guerrilla figures (DAG) include the fighters and civilians involved in skirmishes. Figures for 1946 are from GES (1971) as follows:

January–June 1946, monthly aggregates of all battles and skirmishes, classified by the author.

Figures for the period July–December 1946 are from Tables pp. 54, 87, 93, 99, 110, 158, and 165. Data from GES (1980) are per military operation with the following adjustments: aggregate data for January and February 1947 are split equally per month; for Operation Korax (pp. 173, 179) figures split into May and June; for Operation Lelaps (p. 257) into August and September.

Figures for surrendered guerrillas in 1946 are from GES (1971, Table VII, p. 235). For 1947, data are from GES (1980) for the areas under the A and B Army Corps, and GES (1976) for the areas under the C Army Corps. For 1947, data are from GES (1980), Diagrams 4 and 5, pp. 386–387.

Data for the period 1948–1949 are from US military archives: JUSMAPG History, 1948, 1949.

11.4 Discontent Data

Two sets of data may provide a measure of the political discontent prevailing in post-war Greece: one is the abstention from voting in the March 1946 parliamentary elections following the decision of KKE to boycott the process as allegedly rigged,

Table 11.3 DAG regional formations and morphology

	HQ	Mount height average	Border distance	Guerrilla forces (in persons)							
		(m)	(km)	Jan 48	Mar 49	Apr 49	May 49	Jun 49	July 49	Sep 49	Dec 49
1	Evia	1743	350	120	160	150	150	150	150	130	24
2	Parnas	2292	300	1600	210	650	720	280	100	50	42
3	Othris	1726	250	400	110	200	50	250	110	40	64
4	Agrafa	2069	200	1600	2750	1460	1100	400	260	200	0
5	Pindos	1994	50	1500	250	400	650	430	330	390	34
6	Hasia	1626	150	550	800	1030	550	70	40	120	20
7	Olympos	2084	150	700	1140	550	1300	360	500	270	92
8	Vermion	2014	100	550	300	350	350	500	340	230	55
9	Grammos	2520	1	5500	1450	4000	4300	4800	4850	0	0
10	Vitsi	2128	30	2500	7500	6000	5600	7100	7400	0	0
11	Kaimktsalan	2087	50	1500	860	830	900	780	350	80	20
12	Belles	1445	1	600	750	730	1530	1430	1430	0	80
13	Halkidiki	1404	100	630	340	160	150	160	160	260	100
14	Serres	2212	1	600	1,000	1500	30	30	0	0	0
15	Haidu	1823	1	600	390	270	70	460	450	50	80
16	Thrace	1435	50	2600	950	1,200	600	190	150	180	100
17	Lesvos	967	300		20	20	20	20	20	11	7
18	Samos	1434	500		170	170	150	145	140	74	48
19	Hania	2452	800		40	40	40	40	20	11	7
20	Cefalonia	1628	400		20	20	20	20	20	10	7
21	Peloponnese	1981	400	800	250	60	40	90	80	50	31
	TOTAL			21,550	19,210	19,730	18,280	17,615	16,820	2106	780

Note: For definitions of variables and data sources see Appendix

Table 11.4 Repression and inclusion in interwar Europe

	Civil War	Trade-unionism	School-years	Year of reference
ROM	1907	0.17	1.10	1905
FIN	1918	0.54	2.10	1920
EST	1918	0.68	2.10	1920
GRE	1945–1949	1.16	3.70	1940
HUN	1919	1.62	5.40	1920
POL	1942–1948	4.42[a]	3.00	1940
IRE	1922–1923	4.92	4.70	1920
SPA	1936–1939	7.60	4.30	1935
ITA	1943–1944	8.71	4.50	1940
BEL	1944	9.94	5.80	1940
FRA		2.97	6.43	1930–1940
AUS		10.64	6.67	::
CZE		4.56	7.03	::
NLD		9.13	7.20	::
UK		8.47	7.67	::
GER		9.28	7.80	::
SWE		10.72	8.03	::

Sources: Data for education from Foldvari and van Leeuven (2014, Table 8). For countries with a civil war, the reference year is the one closer to the outbreak of conflict; for those without entries are simple averages of school-years reported for 1930, 1935 and 1940
Data for Trade union participation from Berg-Schlosser and Mitchell (2002, Table 5.3, variable UNION-POP)
[a]The share for Poland is based on 1921 data, but is unlikely to have improved afterwards

(series ABS46); the other is the vote registered against the restoration of Monarchy in the September 1946 Referendum (series REP46).

11.5 Variables

Note	If not stated otherwise, sources are as described in the previous subsections.
DAG	Number of guerrilla fighters with the Democratic Army of Greece, monthly.
DHQ area	Number of DAG fighters grouped per headquarter and mountain formation. *Source:* Data for January 1948, January to July 1949, September and December 1949 are from JUSMAPG, History, 1948–1949, maps.
DISPL	Total number of evacuees from villages in Northern Greece. *Source:* Laiou (1987, Table II, 2. Displaced persons in Northern Greece, per prefecture).
DISTNB	Distance of DAG HQs from northern borders, in km; calculated by the author.

Table 11.5 Chronology for the Greek Civil War

World	Greece	Event
	Aug 1922	Greece defeated in Asia Minor campaign
	Apr 1932	Greece exits the Gold Exchange Standard and defaults
	Aug 1936	Democracy suspended, KKE is banned
	Oct 1940	Italy invades Greece Italian forces defeated
	Apr 1941	Germany invades and occupies Greece
	Oct 1944	Liberation of Greece from German Occupation
	Dec 1944	Violent riots in Athens. British army controls Athens
	Feb 1945	The Varkiza Agreement for disarmament of guerrillas
Feb 1945		The Yalta Conference between US, USSR and UK
May 1945		WW2 ends
	Mar 1946	General elections in Greece KKE abstains
	July 1946	Establishment of emergency martial courts
	Sep 1946	Greek Referendum on the return of the King
Mar 1947		Truman Doctrine proclaimed in US
	Jan 1948	JUSMAG arrives in Greece
June 1949		Yugoslavia breaks ties with USSR
	Aug 1949	Guerrillas are defeated and KKE is banned
March 1957		The Treaty of Rome creates the European Community
	Apr 1967	Military dictatorship in Greece
	July 1974	Democracy restored in Greece. KKE legalised

GNA Number of personnel in the Greek National Army, monthly figures.

MOUNT Altitude of mountains, compiled by the author. Source: http://en.wikipedia.org/wiki/List_of_mountains_in_Greece

PROSEC Number of leftwing citizens prosecuted by emergency martial courts during 1946–1951. Initially martial courts were established in 11 cities, but as the civil war was intensifying they were extended to 30 areas. Allocation is similar to that used for persecutions in 1945–1946. *Source:* Michiotis (2007, Tables 1 & 2, pp. 235–236).

PURGE Number of persons persecuted and victimized during 1945–1946, as described in DAG (1947). Data cover seven areas close to guerrilla HQs in Central and Northern Greece, as explained in Grammenos and Rodakis (1987, pp. 383–390). To correspond the data to more disaggregated guerrilla formations the seven regions are artificially split as shown in Table 5. *Source*: Rizospastis (2011, pp. 138–140).

RCAP Guerrillas captured, monthly.

RKLD Battle-deaths of DAG, monthly.

RSUR Guerrillas surrendered to GNA, monthly.

RWND Wounded guerrillas, monthly.

SKLD Battle-deaths of GNA, monthly.

SMIA GNA soldiers missing in action, monthly.

SWND Wounded of GNA, monthly.